Winterthur Portfolio 10

# Winterthur Portfolio 10

Edited by Ian M. G. Quimby

Published for

The Henry Francis du Pont Winterthur Museum

by the University Press of Virginia

Charlottesville

Statement of Editorial Policy

The objective of The Henry Francis du Pont
Winterthur Museum in publishing *Winterthur
Portfolio* is to make available to the serious student an
authoritative reference for the investigation and
documentation of early American culture.

The publication presents articles about many
aspects of American life. Included are studies that
extend current information about objects used
in America in the seventeenth, eighteenth, and
nineteenth centuries; or about the makers, the
manufacture, the distribution, the use, and the settings
of such objects. Scholarly articles contributing to
the knowledge of America's social, cultural, political,
military, and religious heritage, as well as those
offering new approaches or interpretations concerning
research and conservation of art objects, are
welcome.

Ian M. G. Quimby, *Editor*
Polly Anne Earl, *Associate Editor*

# Contents

# Preface

For the fourth time in its history *Winterthur Portfolio* offers a selection of articles grouped around a central theme or special subject area. *Portfolio 1* dealt largely with the history and development of the Winterthur Museum and its programs. *Portfolio 5* offered articles dealing with the history and arts of Maryland. The articles in *Portfolio 8* treated aspects of religion in American history and particularly the manifestation of religion in architecture and the decorative arts. In *Portfolio 10* we are pleased to present eight articles examining the fine and decorative arts in nineteenth-century America.

*Portfolio 10* was conceived several years ago as an issue devoted to the arts of the Victorian period in America. But what in an American context are the temporal limits of such a period, and is the term meaningful for America at all? The dates of Victoria's reign have no particular significance for Americans, and it seems awkward to lump together under one aesthetic tag ebullient rococo revival with the Arts and Crafts Movement. A more practical approach is the one devised by the Metropolitan Museum of Art in the 1970 exhibition "19th-Century America." That is, let the entire century be treated as a historical unit. Arbitrary, of course, but it has the virtue of simplicity and avoids the controversy inherent in art history terminology. The years from 1800 to 1900 have a special significance for Americans. The year of Thomas Jefferson's election to the presidency was one of deep political division and serious doubt in many quarters over the survival of the young republic. One hundred years later the nation was a continental power with imperial ambitions. American society, not to mention the entire Western world, was about to undergo profound changes; and whether one chooses 1900 or 1914 as a terminal date, the nineteenth century and all we associate with it was undergoing profound transition if not actual dissolution.

The term "Victorian monstrosities" is a familiar one applied to nineteenth-century architecture. Brendan Gill, the indefatigable proselytizer for The Victorian Society in America (founded in 1966), delights VSA members by his rousing lectures in which he never fails to point out that our nineteenth-century architectural heritage is infinitely richer, more varied, more original, and more distinctly American in character than most architecture surviving from the colonial period. For those steeped in the glories of colonial culture, the assertion may come as a shock. Of course there is plenty of vulgarity and ostentation, but serious students of eighteenth-century art are aware that these qualities abounded in that age as well. We tend to select from the past whatever serves present needs.

American art of the nineteenth century was for years beneath the dignity of art historians. Now a new generation of young art historians is busily absorbed in assimilating heretofore neglected and unsuspected riches. The decorative arts have fared even worse, if that is possible, and even "19-Century America" evoked unsympathetic comment from critics who refused to judge it on its own terms. (Jay Cantor's rejoinder to the critics appeared in *Winterthur Portfolio 7*.) Yet that exhibition, whatever its faults, stands as a landmark in the recognition of a major portion of our cultural heritage. The critics notwithstanding, a healthy reappraisal of nineteenth-century American art is now in progress. The articles in *Winterthur Portfolio 10* are another contribution to our understanding of that misunderstood age.

Charles Eastlake's *Hints on Household Taste* was published in America in 1872, four years after its first appearance in England. The popularity of the book in this country is suggested by the fact that it had gone through six American editions, compared with four English editions, by 1879. Eastlake propounded a philosophy of home furnishings that included emphasis on functionalism and integrity of materials and workmanship. Mary Jean Smith Madigan explores the effect of this popular book on American furniture from 1870 to 1890.

Another aspect of the reform movement in nineteenth-century decorative arts is analyzed by Wilson H. Faude in his study of Associated Artists, the decorating firm created in 1879 by the partnership of Louis Comfort Tiffany, Samuel Colman, Lockwood de Forest, and Candace Thurber Wheeler. This formidable combination of talent created unified interior designs of great originality and beauty. The recently restored interior of the Samuel Clemens house (The Mark Twain Memorial) in Hartford, Connecticut, of which Mr. Faude is curator, is a monument to their creative genius.

Associated Artists consciously tried to create an American art in the realm of interior design. The partners thought Americans should not be required to depend on Europe for fine textiles and wallpapers. It is a measure of American technological progress that in the 1880s Associated Artists could command such excellent results. During the first half of the century the country lacked the capacity to produce these products. Caroline Sloat's study of the Dover Manufacturing Company and its laborious efforts to develop techniques for printing designs on cotton textiles is a startling reminder that in the 1820s the United States was still struggling to meet its basic needs in textiles. Fine textiles came from Europe, and this would remain the case throughout much of the nineteenth century.

It was not only the fine fabrics that came from Europe; so also did most of the designs showing their application. Samuel J. Dornsife has provided a visual essay consisting mostly of French and English plates showing recommended window treatments throughout the nineteenth century. These engraved plates constitute the best primary source for documentation of period window hangings. The Dornsife collection of design books on the use of fabrics in nineteenth-century interiors is one of the most extensive collections of its kind. The collection has been donated to The Victorian Society in America.

Gothic revival architecture is for all practical purposes a visual hallmark of the nineteenth century. Of course it was only one of several modes of architectural expression, but perhaps none was more deeply rooted in Anglo-American cultural traditions. Professors Alice Kenney and Leslie Workman in a provocative essay trace the origins and development of the concept of the Middle Ages in Anglo-American culture from 1750 to 1840. They give particular attention to the interplay between the Gothic revival as a literary movement and its effect on architecture and the decorative arts. They propose the interesting thesis that the novels of Sir Walter Scott, which were extremely popular in America, conditioned the acceptance of Gothic revival styles in republican America.

Alexander Turney Stewart and his opulent Fifth Avenue mansion represent a far different aspect of nineteenth-century America. Stewart's rise from immigrant Scots shopkeeper to America's first department store magnate is an astounding success story. In matters of taste he was cautious by standards of the day, usually relying on the advice of others and purchasing the "safe" works of academic painters. But as architectural patron he was bold—sometimes even innovative. Jay Cantor discusses Stewart's architectural achievements against the background of the ever changing architectural scene of mid-nineteenth-century New York City.

Richly ornamented Renaissance revival furniture was the prevailing fashion of the 1860s and 1870s. Rich men like A. T. Stewart could afford to purchase expensive handcrafted editions, but the middle class relied on the products of furniture factories. The eastern cities still dominated the furniture trade, but during the 1870s Grand Rapids emerged dramatically as a center of furniture production. Kenneth L. Ames discusses the development of the Grand Rapids furniture industry and analyzes its products from the standpoint of the art historian.

John La Farge is probably the most undervalued American artist of the nineteenth century. Although born in this country, he was raised by his French émigré parents in the European cultivated tradition. He could easily have been a mere aesthete, considering the age and his circumstances. Instead he became a serious artist who revived in himself the Renaissance concept of the artist as workman. He was a man of great intellect and sensitivity, and one is inclined to think of him as the painter's equivalent of Henry James—who was indeed his close friend. Professor Weinberg's paper discusses the mural paintings executed by La Farge in three Episcopal churches during the 1880s in New York City.

# The Influence of Charles Locke Eastlake on American Furniture Manufacture, 1870–90

*Mary Jean Smith Madigan*

WHEN CHARLES LOCKE EASTLAKE'S trend-setting book on the reform of interior design and the manufacture of domestic articles, *Hints on Household Taste,* was first published in the United States, in 1872, it excited a flurry of interest and controversy that reached its peak in the late 1870s, and it continued to affect the appearance of American interiors for years after that. Intended to instruct the masses in the principles of tasteful home decoration, Eastlake's volume was influenced by the design reform movement in England. It expressed his view that furniture should be functional, nonostentatious, simple and rectilinear in form, honestly constructed "without sham or pretense," and ornamented with respect for the intrinsic qualities of the wood as well as the intended uses of the furniture. Eastlake's illustrations of well-designed furniture, which included several of his own sketches, showed a number of ornamental features including shallow carving, marquetry, incised or pierced geometric designs, rows of turned spindles, brass strap hinges, bail handles, and keyhole escutcheons inspired by medieval forms, all used in decorative but functional ways.

His views aroused controversy among American critics of taste, whose divergent opinions set the stage for continuing confusion about Eastlake's role and the extent of his influence on American furniture styles. Harriet Prescott Spofford, the well-known novelist, noted in 1878 that "a number of essays, written by Mr. C. L. Eastlake, were printed in various English publications, and afterwards collected in a volume that has done great work toward revolutionizing the manufacture of furniture . . . the book met a great want. Not a marrying couple who read English were to be found without *Hints on Household Taste* in their hands, and all

its dicta were accepted as gospel truths." In that same year another influential critic, Clarence Chatham Cook, wrote: "The 'Eastlake' furniture must not, however, be judged by what is made in this country, and sold under that name. I have seen very few pieces of this that were either very well designed or well made. None of the cheaper sort is ever either." [1]

Accounts of the development of late nineteenth-century American furniture styles that make reference to Eastlake usually do so with condescension toward a would-be revolutionary whose "well-meaning ineptitudes" were misinterpreted in manufacture.[2] The nature of the criticism directed against Eastlake and Eastlake furniture by American commentators centers around three issues: whether a definable "Eastlake" style existed in America between 1870 and 1890; whether this style or styles conformed in manufacture to Eastlake's dicta on construction, function, and ornamentation; and whether it was economically feasible in America to mass-produce and distribute at low cost furniture true to Eastlake's tenets. Was American "Eastlake" furniture a debased travesty of his designs and the philosophical spirit of his *Hints on Household Taste,* as some critics have suggested, or did it constitute a true revolution in popular taste and a prelude to functionalism in design? In at-

[1] Spofford, *Art Decoration Applied to Furniture* (New York: Harper & Bros., 1878), p. 147; Cook, *The House Beautiful: Essays on Beds and Tables, Stools and Candlesticks* (New York: Scribner, Armstrong & Co., 1877), p. 223.

[2] For example, see Joseph Aronson, *The Book of Furniture and Decoration* (New York: Crown Publishers, 1936), p. 121; Marta Sironen, *A History of American Furniture* (East Stroudsburg, Pa.: Towse Publishing Co., 1927), p. 51; Felicia Davis, "The Victorians and Their Furniture," *Antiques* 43, no. 6 (June 1943): 258; Alan Gowans, *Images of American Living* (New York: J. B. Lippincott, 1964), p. 336.

tempting to answer this question, the three points of criticism mentioned above will be evaluated in terms of the furniture produced in America during the 1870s and 1880s. But first it may be helpful to consider the spirit of English design reform that inspired *Hints on Household Taste,* as well as Eastlake's background, the philosophy he expressed, and the dissemination of his ideas in the United States.

Mid-nineteenth-century England was preoccupied with the search for a national style of art and architecture. Augustus Welby Northmore Pugin (1812–52) and John Ruskin (1819–1900) pressed for revival of the "honest" principles of Gothic design and architectural construction. They were opposed to the English classicists, who favored a style of architecture influenced by Renaissance forms. This architectural "battle of the styles," which raged for two generations, had its counterpart in the decorative arts as well. At midcentury, manufacturers of furniture in both England and America relied heavily for inspiration on design elements of the Renaissance and rococo periods. The state of common manufacture was exemplified by a number of tastelessly florid and ostentatious pieces displayed at the Great Exhibition of 1851 in London's Crystal Palace. Reacting to these misapplications of a foreign design aesthetic, reformers such as William Morris (1834–96) championed a return to the principles of English medieval craftsmanship in furniture and other decorative arts. In 1861 Morris and a group of friends including Philip Webb, Ford Madox Brown, Dante Gabriel Rossetti, P. P. Marshall, and C. J. Faulkner founded a decorating firm that turned out an "artistic" line of handcrafted furniture, stained glass, and tapestries inspired by medieval sources. The two Morris, Marshall, Faulkner and Company entries in the 1862 International Exhibition at South Kensington encouraged a trend for medieval furniture design, which was also expressed in the work of William Burges (1827–81) and Bruce Talbert (1838–81). In 1867 Talbert, a Scottish architect who designed furniture for Gillow and Company, Manchester, published a book of thirty plates of furniture designs entitled *Gothic Forms Applied to Furniture, Metal Work, and Decoration for Domestic Purposes.* The book met with great success and was reprinted in the United States in 1873 and 1877.[3]

The clamor over Gothic revivalism and the art movement, as well as the awakening spirit of reform in English domestic manufacture, influenced the thought and writings of Charles Locke Eastlake. Born March 11, 1836, at Plymouth, Devonshire, to a family prominent in English art circles, Eastlake was well prepared for his later role as an arbiter of taste. He was guided and educated by his uncle, Sir Charles Lock Eastlake, president of the Royal Academy, who was keeper and later director of the National Gallery. A talented draftsman, the young Eastlake was trained as an architect, although he never executed a building. As a young man he traveled extensively on the Continent and wrote articles on architecture and art for popular journals. During his later tenure as secretary of the Royal Institute of British Architects (1855–78), Eastlake published his two best-known works, *Hints on Household Taste* (1868) and *A History of the Gothic Revival* (1872). Through the influence of his aunt, Lady Elizabeth Rigby Eastlake, the widow of Sir Charles, he was appointed in 1878 to the keepership of the National Gallery. There, Eastlake classified the enormous painting collections and introduced needed conservation measures. He retired in 1898 to a life of simple domestic pleasures at his Terra Cottage in Bayswater, where he died in 1906.[4]

During his administrative career, Eastlake published a prodigious number of articles on painting, architecture, and taste; delivered a lecture series; and wrote two major books reflecting his preoccupation with the medieval revival, design reform, and domestic comfort. His most scholarly work, *A History of the Gothic Revival,* which as "a standard textbook on Victorian gothic has never been superseded," describes and documents the design origins of 343 examples of neo-Gothic buildings executed in England between 1820 and 1870. Eastlake's assessments of various revivalist writers and architects exposed his predilection to Ruskinian philosophy. He described Ruskin's *Stones of Venice* as having "no parallel in English literature." Although he lamented Ruskin's "unhappy alliance" of art criticism and utopian social philosophy, there is little doubt these ideas helped to form Eastlake's own opinions on the moral qualities of honest craftsmanship.[5]

---

[3] Talbert, *Gothic Forms Applied to Furniture, Metal Work, and Decoration for Domestic Purposes* (1867; reprint ed., Boston: J. R. Osgood & Co., 1873).

[4] For biographical data, see *The Builder* 91 (1906): 607; J. Mordaunt Crook, Introduction to Charles L. Eastlake, *A History of the Gothic Revival* (1872; reprint ed., New York: Humanities Press, 1970), pp. 18–23.

[5] Crook, Introduction to Eastlake, *History of the Gothic Revival,* p. 5; Eastlake, *History of the Gothic Revival,* p. 105.

The best remembered of his works is *Hints on Household Taste.* First published in 1868 from a series of articles that had appeared earlier in *The Queen* and *London Review,* this book was printed four times in England and six in America.[6] Unlike Talbert's *Gothic Forms Applied to Furniture,* Eastlake's work was not a sketchbook of designs. Although it contained a number of furniture illustrations, some designed by Eastlake himself, *Hints on Household Taste* had a much broader aim: to suggest "some fixed principles of taste for the popular guidance of those who are not accustomed to hear such principles defined."[7] The book was inexpensively bound for mass distribution. Because *Hints on Household Taste* has been recently reprinted, a detailed résumé is unnecessary, but it may be helpful to reiterate its general tenets, with special consideration given to those points that are most often overlooked or misinterpreted.

In *Hints on Household Taste* Eastlake deplored the overornamentation and shoddy ostentation that characterized much of English furniture and many other articles of contemporary domestic manufacture. Like Ruskin, he blamed this degenerate taste on the separation of art from manufacture. "We can hardly hope, then, in our own time, to sustain anything like a real interest in art while we tamely submit to the ugliness of modern manufacture. We cannot consistently have one taste for the drawing room and another for the studio." He urged recognition of certain definable principles of function and appropriateness common to good taste in both structural science and fine art; these principles were to be "derived from common sense." Rejecting the prevalent idea that good taste was instinctive, "the peculiar inheritance of gentle blood, and independent of all training,"[8] Eastlake pointed out that by default shopkeepers, upholsterers, and milliners—individuals motivated by profit and untrained in aesthetics—had become the arbiters of taste for the nation.

Eastlake recognized and lamented the operation of economic factors that made "novelty" desirable to producer and consumer alike and that encouraged extravagance of design and "a show of finish" over more enduring qualities of good craftsmanship. Nevertheless, he accepted the inevitability of mass production of household goods by machine processes, a fact that is overlooked by his detractors. Although he immensely admired the wholly handcrafted furniture of earlier times, as exemplified in his drawing of a 1620 sofa that appeared in *Hints on Household Taste* (Fig. 1), Eastlake did not ad-

FIG. 1. Charles L. Eastlake, drawing of 1620 settee in billiard room at Knole House, Kent, England. From Eastlake, *Hints on Household Taste,* ed. Charles C. Perkins (Boston: James R. Osgood & Co., 1872), pl. 29, opp. p. 171.

vocate the abandonment of machine processes for a return to hand labor. Acknowledging that "human handiwork . . . will always be more interesting than the result of mechanical precision," he admitted that "division of labour and perfection of machinery have had their attendant advantages, and it cannot be denied that many articles of ancient luxury are by such aid now placed within reach of the million. . . . it would be undesirable, and indeed impossible, to reject in manufacture the appliances of modern science."[9]

It was not mass production that he deplored, but the misapplication of machine technology to create a surfeit of tasteless ornamentation—"wood-mould-

[6] Eastlake, *Hints on Household Taste* (London: Longmans, Green & Co., 1868; 2d ed., 1869; 3d ed., 1872; 4th ed. rev., 1878); Eastlake, *Hints,* ed. Charles C. Perkins (Boston: J. R. Osgood & Co., 1872; 2d ed., 1874; 3d ed., 1875; 4th ed., 1876; 5th ed., 1877; Boston: Houghton Mifflin Co., 1879). The fourth revised British edition (1878) was reprinted in New York by Dover Publications in 1969. Unless otherwise noted, subsequent references are to this reprint edition.

[7] Eastlake, *Hints* (Boston: J. R. Osgood & Co., 1872), p. xxiv.

[8] Eastlake, *Hints,* pp. 7, 1, 8. "Drawing room" in the sense Eastlake used it refers to the design room, where mechanical "drawings" for production were made.

[9] Eastlake, *Hints,* pp. 8, 3, 105, 106. On p. 134, Eastlake describes a "picturesque" library table that he finds pleasing despite its "inevitable faults of modern joinery, viz., adhesive mouldings, 'mitred' joints, etc."

ings . . . by the yard, leaf-brackets by the dozen, and 'scroll-work' . . . by the pound"—which was then applied, indiscriminately and nonfunctionally, to furniture. Eastlake's philosophy of functionalism (though he did not so name it) was basic to his definition of the principles of good taste. "Every article of manufacture which is capable of decorative treatment should indicate, by its general design, the purpose to which it will be applied. . . . particular shapes and special modes of decoration are best suited to certain materials. Therefore the character, situation, and extent of ornament should depend on the nature of the material employed, as well as on the use of the article itself." [10]

Furniture was tasteful only when it conformed to the constructional and decorative limitations imposed by the wood itself, "chaste and sober in design, never running into extravagant contour or unnecessary curves." Curved or shaped furniture, especially that of the mid-Victorian rococo revival, was particularly offensive. "The tendency of the last age of upholstery was to run into curves. Chairs were invariably curved in such a manner as to ensure the greatest amount of ugliness with the least possible comfort. The backs of sideboards were curved in the most senseless and extravagant manner; the legs of cabinets were curved, and became in consequence constructively weak; drawing-room tables were curved in every direction, . . . and were therefore inconvenient to sit at, and always rickety. This detestable system of ornamentation is called 'shaping.'" [11] It should be emphasized, however, that Eastlake was not a confirmed advocate of any particular style of furniture. Although the designs illustrated in his book show decided medieval proclivities, Eastlake recognized and accepted the eclectic nature of furniture design in the 1860s and 1870s and had no wish to impose his personal stylistic preferences on his public.

He preferred simplicity to extravagance, but where "an effect of richness" was desired, Eastlake advocated the judicious use of carving, inlay, and sometimes even veneer in furniture decoration. Shallow carving was permissible where its presence did not interfere with comfort or utility and when it typified or symbolized, but never literally repro-

duced, natural forms, for "decorative art is degraded when it passes into a direct imitation of natural objects." Eastlake did not approve of carved ornamentation on "second-rate" furniture, for such machine-produced decoration was often glued on inappropriate places in a manner "egregiously and utterly bad." In his view applied carving was better eliminated altogether. Other legitimate modes of decoration included marquetry, inlaid woodwork, and, in certain cases, veneering. Although he deplored the "sham" of making one material look like another, he was realistic about the common practice of veneering. "If we are to tolerate the marble lining of a brick wall and the practice of silver-plating goods of baser metal,—now too universally recognized to be considered in the light of a deception—I do not see exactly how veneering is to be rejected on moral grounds." [12]

Like other English reform designers, Eastlake favored "honesty" in principles of construction. In architecture as well as furniture manufacture, he deplored covering shoddy workmanship with inappropriate ornamentation, for "the first principles of decorative art . . . require that the nature of construction, *so far as is possible,* should always be revealed . . . by the ornament which it bears." Nowhere did Eastlake dictate that hand-mortised-and-doweled joints, though desirable, were the only permissible recourse of the maker, nor did he require that every piece of furniture have its "flayed skeleton" exposed to public view for the sake of honesty in construction. In the preface to the first American edition of *Hints on Household Taste* he reiterated that "it is the spirit and principles of early manufacture which I desire to see revived, and not the absolute forms in which they found embodiment." Eastlake considered the country cartwright a good exemplar of these "spirits and principles of early manufacture," for "his system of construction is always sound, and such little decoration as he is enabled to introduce never seems inappropriate, because it is in accordance with the traditional development of original and necessary forms." [13] Applying his own principles in furniture designs, Eastlake used turned members for decora-

---

[10] Eastlake, *Hints,* pp. 58, 92. On p. 172, Eastlake states, "In the sphere of what is called industrial art, use and beauty are . . . closely associated: . . . the humblest article of manufacture, when honestly designed, [has] a picturesque interest of its own."

[11] Eastlake, *Hints,* pp. 159, 55–56.

[12] Eastlake, *Hints,* pp. 68, 69, 58, 59. He particularly deplored the use of naturalistic carved trophy swags of "slaughtered hares and partridges" on Renaissance revival sideboards.

[13] Eastlake, *Hints,* p. 109, italics added; Eastlake, *Hints* (Boston: J. R. Osgood & Co., 1872), p. xxv; Eastlake, *Hints,* p. 61.

FIG. 2. Charles L. Eastlake, design for a sideboard. From Eastlake, *Hints on Household Taste,* ed. Charles C. Perkins (Boston: James R. Osgood & Co., 1872), pl. 12, opp. p. 85.

tive supports to good advantage, as in the dining room sideboard illustrated in *Hints on Household Taste* (Fig. 2).

Like other writers on taste in the late nineteenth century, Eastlake ran headlong into the dilemma of the economics of supply and demand, which often kept well-designed goods out of the marketplace. He felt that tasteful furniture "ought really to be as cheap as that which is ugly," since "the draughtsman and mechanic must be paid, whatever the nature of their tastes may be." He noted that shopkeepers charged exorbitant prices for any desirable novelty of design, and tasteful, rectilinear art furniture was considered a novel departure

from the curving rococo monstrosities in current production. Eastlake knew it would be "quixotic to expect any one but a wealthy enthusiast to pay twice as much as his neighbor for chairs and tables in the cause of art," but he saw no reason why well-designed furniture could not be cheaply mass-produced, for "the true principles of good design are universally applicable, and if they are worth anything, can be brought to bear on all sorts and conditions of manufacture." [14] In sum, there was little in Eastlake's book to preclude the application of his ideas on furniture design, construction, and manufacture to the system of mass production of furniture then developing in America.

Furniture manufacture in the United States in the several years preceding 1870 was undergoing a transition in scale. Census figures indicate that the number of workmen employed in the industry more than doubled between 1860 and 1870, an increase not directly attributable to the much smaller comparative rise in population. Most of the increased output is traceable to the newer factories in towns west of Boston and New York, the traditional furniture centers: Jamestown, New York; Williamsport, Pennsylvania; Cincinnati and Cleveland, Ohio; Grand Rapids, Michigan; and Indianapolis, Indiana, among others. Equipped with the latest machinery (often steam-powered) and amply supplied with native timber (cherry, ash, walnut, maple, and oak), the new factories were located on rail lines to facilitate shipment of their products to the eastern shopkeepers who conducted the retail trade. Small eastern producers, especially cabinetry shops, felt the squeeze of competition. Ernest Hagen, a New York cabinetmaker, recalled that shortly after 1867 "the great change came over the cabinet making trade of New York. The factory work, and especially the Western factory work drove everything else out of the market." The new factories, with scroll and circular saws and mortising and carving machines, "looked for a fashion in which they could use their facilities to best advantage . . . this they found in the Renaissance, which for a number of years superseded all other styles in the best class of furniture." [15]

[14] Eastlake, *Hints,* pp. 174–75, 91.

[15] Hagen quoted in Elizabeth A. Ingerman, "Personal Experiences of an Old New York Cabinetmaker," *Antiques* 84, no. 5 (Nov. 1963): 580; George W. Gay, "The Furniture Industry in America," *Encyclopedia Americana* (1909). Gay, himself a manufacturer, was associated with the Berkey and Gay Furniture Company, Grand Rapids, Michigan.

In the years following the Civil War, while American energies were redirected from military endeavor to the creation of a new industrial and social order, American homes were being filled with heroic monuments to historicism—massive walnut "Renaissance" furniture surmounted by broken pediments and heavy finials, overdecorated with applied ornamentation in the form of acanthus leaves, scrolls, medallions, swags, cartouches, and all manner of molding. But as Americans increasingly traveled abroad or read English periodicals, an awareness of the English movement for design reform filtered into their aesthetic consciousness. Affected by the pervasive Victorian drive toward self-improvement and released from the more pressing preoccupation of the war, middle-class Americans provided an eager audience for the professors of taste, of whom Charles Locke Eastlake was the foremost.

While Eastlake's ideas were politely received in England, his overall influence was greater in the United States, as evidenced by the much stronger demand there for *Hints on Household Taste*. In his homeland, Eastlake's was one of many voices crying out for design reform, whereas in America his words assumed the special authority inevitably accorded European arbiters of taste by culturally self-conscious Americans who sought reassurance in aesthetic dictates from the established "Old World." The popularization of Eastlake's ideas in the United States began in 1872, with the publication of the first American edition of *Hints on Household Taste,* edited and extensively prefaced by Charles Callahan Perkins, a cultured Bostonian remembered chiefly for his books on Italian sculpture. *Hints on Household Taste* was not new to those affluent Americans who had had access to the three earlier British editions or who had read critical reviews in the *Art Journal,* a periodical published by Appleton's in the 1870s. As Perkins noted, "The leading periodicals and newspapers of the United States have generally commended the present work to all who desire information upon the subject of which it treats." Nevertheless, the publication of an American edition brought Eastlake's ideas to unprecedented numbers of new readers, and, by 1874, Eastlakean tenets of simplicity and taste were being implemented in home furnishings. The *New York Times* noted that "people are becoming more simple and austere in the manner of house furniture. . . . upholsterers and furniture dealers do say, with some show of regret, that the demand for extravagant and florid goods for household use is gone."[16]

Furniture makers, uncertain of the effect of Eastlake's ideas on established methods of manufacture, reacted with more panic than logic. A writer in the *Cincinnati Trade List* declared: "We hope the fashion won't live very long or extend to this city. Of all the clumsy, ugly inventions, or rather copies, the sort advocated with bigoted zeal by Eastlake deserves to be most condemned."[17] Despite the protest from Cincinnati and from others who feared that Eastlake's call for a return to simplicity would destroy the furniture industry, the influence of *Hints on Household Taste* was growing. It was extended not only by five reprintings between 1873 and 1890 but also by the echo effect of American authors who repeated, sometimes verbatim, Eastlake's major themes of honesty, simplicity, functionalism, and appropriateness, as well as his specific recommendations on ornament and decoration. Articles and books by these authors, with illustrations of tasteful furniture similar to those appearing in *Hints on Household Taste,* must be credited for the thorough dissemination of Eastlake's gospel and for creating a demand for furniture that conformed to his principles, although in a variety of derivative styles.

One of the most important of Eastlake's American popularizers was the architect Henry Hudson Holly (1843–92), who, in a series of articles for *Harper's Monthly,* repeated entire passages from *Hints on Household Taste* without giving credit to the author. Holly echoed Eastlake's opinions on wall decoration, floor and window treatments, metalwork, shaped furniture, bookcases, bedroom furniture, and the general venality of shopkeepers. He acknowledged Eastlake as the source of his ideas only once, in a paragraph condemning pedestal tables: "Mr. Eastlake, with justice, we think, condemns these rattletraps altogether as unconstructional." One of Holly's articles did include an illustration of a bedroom suite that he described as "a group of bedroom furniture of medieval design, commonly known in this country as Eastlake style, which recommends itself by its simplicity and hon-

---

[16] Perkins, Preface to Eastlake, *Hints* (Boston: J. R. Osgood & Co., 1872), p. v; *New York Times,* Oct. 5, 1874.

[17] *Cincinnati Trade List* quoted in *Furniture Gazette* (London), vol. 2, May 24, 1874. This quote was brought to my attention by Catherine L. Frangiamore, assistant curator, Cooper-Hewitt Museum of the Smithsonian Institution, New York.

FIG. 3. Henry Hudson Holly, design for a bedroom suite. From Holly, "Modern Dwellings: Their Construction, Decoration, and Furniture," *Harper's Monthly* 53, no. 315 (Aug. 1876): 358, fig. 9.

est treatment" (Fig. 3). To Holly's credit, when his articles were published in book form, in 1878, he eliminated those passages that had been cribbed from *Hints on Household Taste* and noted in the preface that he had "profited considerably by the writings of Eastlake and others." [18]

While the echo effect of Eastlake's *Hints on Household Taste* is most apparent in the writings of Henry Hudson Holly, it can also be found in a series of articles on household art written in 1876 for the *Art Journal* by Charles Wyllys Elliott (1818–83). Elliott, a landscape gardener and student of Andrew Jackson Downing, was a commis-sioner for the laying out of Central Park in New York City. He wrote extensively on historical, literary, and artistic subjects and for a time managed the Household Art Company in Boston, which dealt in tasteful household goods. Although his particular interest was ceramics, Elliott accurately echoed Eastlake's doctrines of good taste in everything from architecture to ornament to furniture construction. He often pinpointed the same examples of "bad taste"—realistic trophy swags on sideboards, curved Louis Quinze chairs, extension tables—that Eastlake had found offensive.[19]

Elliott observed that Americans of 1876 were re-

[18] See Holly, "Modern Dwellings: Their Construction, Decoration, and Furniture," *Harper's Monthly* 53, nos. 314–15 (July–Aug. 1876): 217, 225–26, 357; Eastlake, *Hints*, pp. 8, 11, 55; Holly, *Modern Dwellings in Town and Country* (New York: Harper & Bros., 1878), preface.

[19] Elliott, "Household Art," *Art Journal* 1, nos. 11, 12 (Nov., Dec. 1875): 295–300, 333–36; 2, nos. 13, 14, 16, 19, 22 (Jan., Feb., Apr., July, Oct. 1876), 9–15, 50–55, 116–23, 180–85, 239–44.

FIG. 4. Charles Wyllys Elliott, design for a sideboard. From Elliott, "Household Art," *Art Journal* 2, no. 19 (July 1876): 181, fig. 4. (Art and Architecture Division, New York Public Library, Astor, Lenox, and Tilden foundations.)

volting against "brilliance which was gaudy, excessive ornamentation, and vulgar luxury" in furnishing their homes. As a result, "no man will wish to spend his life with a 'furbelowed' sideboard as he would not with a 'furbelowed' woman." Although Elliott credited Eastlake with "excellent work in rousing attention to the fact that there is such a thing as Household Art," he criticized the furniture illustrated in *Hints on Household Taste* as "far from good" and "too costly." Ironically, Elliott's own furniture illustrations bore a striking resemblance to Eastlake's, as a comparison shows (Figs. 2, 4). Elliott hesitated to ascribe the name "Eastlake" to the kind of "simple, real, strong, and honest" furniture they both advocated. "What style do you call this . . . is it a copy of the medieval, the gothic, the Roman, or what? Is it Eastlake? It is none of these. It is an attempt to express a useful purpose, with fine lines, with modest decoration, and with honest construction. . . . it is not easy to invent a name. . . . if obliged to coin one, I should like to try 'The Homelike Style,' because the main purpose of it is not display, ostentation or luxury, but to help make the home life beautiful." [20]

[20] Elliott, "Household Art," *Art Journal* 2, no. 19 (July 1876): 180–83.

Perhaps the most incisive and analytical writer to echo Eastlake was Harriet Prescott Spofford (1835–1921). Known for her romantic novels, Spofford undertook in 1876 a series of articles for *Harper's* on the history of furniture styles. She had a particular liking for old Gothic furniture and for the new styles it inspired. "Nothing can be better than the gothic for the rich, permanent, abounding appearance due the dining room; it recalls its old feasts and orgies." However, she applied the Eastlakean standard of "simplicity and truth and frankness of construction," to all furniture, crediting Eastlake himself with bringing order out of the chaos of Victorian "ugliness, slovenliness and stupidity." [21]

In 1878, Harriet Spofford's articles were expanded into a book, *Art Decoration Applied to Furniture,* of which one chapter was devoted to "The Eastlake." Here Spofford praised Eastlake's *Hints on Household Taste* as a "volume that has done great work toward revolutionizing the manufacture of furniture." In her articles for *Harper's,* Spofford had helped to spread the Eastlake gospel by repeating his dicta, but in *Art Decoration* she evaluated, as an impersonal commentator, the effect of *Hints on Household Taste.*

The book occasioned a great awakening, questioning, and study in the matter of household furnishings. Presently there arose a demand for furniture in the "Eastlake style." . . . the upholsterers, with whom Mr. Eastlake made quarrel in his pages, denied that there was any such style. . . . The demand, however, was one which obliged the upholsterers to pocket their grudge, and if there were no Eastlake style to invent one, for today Eastlake chairs . . . Eastlake bedsteads, clean shaped and charming, Eastlake washstands, dressing-cases, drawers and cabinets, are to be seen everywhere.

Spofford observed that since Eastlake himself designed so few pieces, makers were forced to invent their own furniture "after Eastlake" to satisfy the public demand. They succeeded in producing "an interesting variety, quaint, with an attractive air of antiquity, full of character . . . but always a little stiff. The upholsterers themselves have no fancy for its straight up and down angularity. . . . although they manufacture the article, they still seem to dislike calling it The Eastlake, and with some reason as it so nearly fulfills the requirements of the medieval as scarcely to need a separate name."

[21] Spofford, "Medieval Furniture," *Harper's Monthly* 53, no. 318 (Nov. 1876): 828.

Finally, Spofford recognized that Eastlake's ultimate contribution to furniture manufacture was not as the designer and creator of a single distinct style but as the prime force in "a movement, seldom, if ever before effected by a single person; [Eastlake] has succeeded in inaugurating a new regime, which bears the same relation to the loose and wanton Quatorze and Quinze regimes that virtue bears to vice." [22]

For frivolous-minded readers without patience or inclination to wade through long articles by Holly, Elliott, and Spofford, there existed yet another Eastlakean disciple, one whose enthusiastic effusions graced the *Art Journal*. She was Mary Elizabeth Wilson Sherwood (1826–1903), daughter of a congressman, wife of a lawyer, sometime novelist, and indefatigable socialite. Her short articles on fashion and interiors, usually signed with her initials, M.E.W.S., gave the stamp of social approval to Eastlake-influenced interiors in affluent homes. Reviewing the "tastefully" decorated town house of F. W. Stevens, Esq., she remarked: "The world seems flooded with archaeology, architecture, and colour; there is an epidemic of splendor. . . . we must bravely dare to be conservative, or else swim with the stream of Ruskin, Eastlake, or Morris." [23]

The most influential of Eastlake's popularizers—who would probably have rejected that designation—was Clarence Chatham Cook (1828–1900). An art critic, editor, and writer for the *New York Times*, Harvard-educated Cook was a cultivated man who wasted little sympathy on those lacking education or means. In 1877 he published *The House Beautiful: Essays on Beds and Tables, Stools and Candlesticks,* a compilation of articles that had previously appeared in *Scribner's Magazine.* Cook joined Eastlake's crusade for simplicity, functionalism, and honest construction in furniture. "I have reached a point where simplicity seems to me a good part of beauty, and utility only beauty in a mask; and I have no prouder nor more pretending aim than to suggest how this truth may be expressed in the furniture and decoration of our homes." Cook echoed Eastlake in specifics as well as in generalities: rugs should not extend to the walls, staining is a desirable furniture finish and varnish an abomination, bookcases do not require doors, and so forth. Even

the illustrations in *The House Beautiful* paralleled those in *Hints on Household Taste.* Apparently unmindful of his debt to Eastlake, Cook vehemently rejected anything labeled "Eastlake," especially American-made furniture. However, he admired those Eastlake pieces that were custom-made by the New York firms of Herter Brothers and Leon Marcotte, but cautioned that "they are only referred to as doing the style (if it be a style) more justice than the lumps of things we see in certain shops, though, in truth, these lumps are a good deal more like the things recommended in Mr. Eastlake's book than the stylish, elegant pieces designed by Messrs. Herter and Marcotte." [24]

To place Cook's allegations in perspective, it must be remembered that he was an elitist who advised his readers to shop at Herter's and Cottier's, deploring factories where "the logs go in at one door, and come out at another fashioned in that remarkable style known here as Eastlake and which has become so much the fashion that grace and elegance are in danger of being *taboo* before long." Cook was a fanatic on certain issues, as well as a bigot. He disliked central heating, gas lighting, and sanitary plumbing facilities; he blamed the widespread American adoption of these conveniences on the laziness of Irish servants who refused to take out slop jars, fill kerosene lamps, and so forth. "Before the invasion of the Biddy tribe from the bogs of Ireland, the labor of the house was looked upon as a part of the price paid for domestic enjoyment. But the Irish . . . bluntly refused to do the work required even if they were paid for it." [25] In the light of Cook's tendency toward elitism and his irrational bias, it seems unfortunate that later writers placed so much reliance on his assessment of American-made Eastlake furniture.

As Eastlake's tenets were being spread by the popular press in America, examples of Eastlake-influenced art furniture of British and American make were displayed at the Philadelphia Centennial Exhibition. In the summer of 1876, thousands of Americans visited this "great exhibition" devoted to the union of art and industry, while still more thousands of stay-at-homes enjoyed the innumerable illustrated registers and commemorative journals published for national circulation. Though eclecticism was characteristic of the Centennial Exhibition, the best of the British furniture

[22] Spofford, *Art Decoration*, pp. 147, 150, 153.
[23] Sherwood, "Some New York Interiors," *Art Journal* 3, no. 36 (1877): 361.

[24] Cook, *House Beautiful*, pp. 22, 224.
[25] Cook, *House Beautiful*, pp. 282, 271.

Fig. 5. Collinson and Locke, London, design for a sideboard exhibited at the Philadelphia Centennial of 1876. From Walter Smith, *The Masterpieces of the Centennial International Exhibition,* 3 vols. (Philadelphia: Gebbie & Barrie, 1876), 2:240.

there reflected Eastlakean precepts of good taste: rectilinear form, relative simplicity of outline and detail, turned balusterlike supports and rows of spindles, and shallow carving or inlaid ornamentation. Cooper and Holt of London presented an ebonized drawing room cabinet with inlaid and tiled panels, while Collinson and Locke showed a buffet (Fig. 5) that one reviewer described as "after the manner made familiar to us in this country by Mr. Eastlake in his book on Household Art." Although this buffet was designed in the so-called Queen Anne revival style, then popular in England, the American commentator saw no incongruity in relating it to Eastlake's precepts. Similar furniture was exhibited by other British manufacturers, including William Scott Morton and Company of Edinburgh and James Schoolbred of Lon-

don, whose firm favored "furniture in the style of the middle ages." [26]

While critics conceded that British furniture was superior to American in both design and "truths of construction," a few American makers showed pieces conforming to Eastlake's standards. Kimbel and Cabus of New York exhibited an entire drawing room, replete with dado, mantle, beamed ceiling, and door cornice, all in ebonized cherry (Fig. 6). "Modern Gothic" in style, the Kimbel and Cabus furniture was "profusely gilt" in ornamentation and featured a sofa with a row of turned spindles at the back. Since the pedestal pictured in their exhibit was also listed in their catalog for 1876, one can assume that Kimbel and Cabus furniture was not all custom-designed and that most of it was available from stock. Mitchell and Rammelsburg of Cincinnati, Ohio, showed a sideboard and a mirrored hallstand designed, according to one reviewer, "rigidly after the canons of Eastlake."[27] Medieval in feeling, made of plain oak with burnished steel hinges, the Mitchell and Rammelsburg pieces were at once functional and ornamental (Fig. 7). They were decorated with simple turnings, chamfered columns, shallow carved panels, and incised work; like the Kimbel and Cabus pedestal, they were apparently taken from available stock.

Most of the American furniture on exhibition at Philadelphia, whether custom- or factory-made, adhered to the stylistic imperatives of the still-popular Renaissance revival. Although custom-made Renaissance revival pieces, like the award-winning sideboard by Giuseppe Ferrari of New York, were skillfully carved and "honestly" constructed, many of the factory-made Renaissance revival pieces sprouted an overabundance of applied ornament that may have precipitated George T. Ferris's criticism that American furniture was "badly designed, tawdry, and vulgar work, only fit at best for the drawing room of a parvenu, or the glittering salon of a north river steam boat." Ferris, whose *Gems of the Centennial* (1877) described and illustrated major exhibition pieces, regretted that there were so

[26] Walter Smith, *The Masterpieces of the Centennial International Exhibition,* 3 vols. (Philadelphia: Gebbie & Barrie, 1876), 2:245; James D. McCabe, ed., *The Illustrated History of the Centennial Exposition* (Philadelphia: National Publishing Co., 1876), p. 365.

[27] George T. Ferris, *Gems of the Centennial Exhibition* (New York: D. Appleton & Co., 1877), pp. 137–38, 141.

FIG. 6. Kimbel and Cabus, display of drawing room fittings and furniture of ebonized cherry at the Philadelphia Centennial Exhibition of 1876. From George T. Ferris, *Gems of the Centennial Exhibition* (New York: D. Appleton & Co., 1877), p. 138.

FIG. 7. Mitchell and Rammelsburg, display of an oak sideboard and hallstand at the Philadelphia Centennial Exhibition of 1876. From George T. Ferris, *Gems of the Centennial Exhibition* (New York: D. Appleton & Co., 1877), p. 141.

few pieces of Eastlake-inspired furniture shown at Philadelphia, a fact he considered "somewhat singular . . . in view of the extraordinary discussion and interest aroused everywhere within the last few years by Mr. Eastlake's studies of Household Art." This critic attributed the absence of Eastlake-inspired pieces to the profit motive of manufacturers, who "do not take kindly to this school of design and decoration, in virtue of the fact that it threatens to result in more simple and less profitable forms of work." [28]

Not only did the Philadelphia Centennial demonstrate that Eastlake's influence had, by 1876, affected a number of American furniture makers; the exhibition itself scattered the seeds of design reform among other manufacturers, particularly in the Middle West. Assessing the effect of the Centennial on American furniture manufacture, George W. Gay, the president of Berkey and Gay, an important early Grand Rapids firm, noted its "far reaching influence, especially on western manufacturers, who until this time had not had occasion to compare their products with those of the best manufacturers of America and Europe. The Eastlake style was quickly taken up by the manufacturers of cheaper furniture, who until then had given very little attention to artistic form." In their zeal to duplicate the look, if not the quality, of Eastlake-influenced pieces shown at the Centennial, it is likely that American manufacturers turned out a good many hybrids, few more tasteful than the hackneyed rococo and Renaissance revival pieces they were intended to replace. Early attempts by western factories to mass-produce Eastlake furniture were clumsy. According to Gay, "an insufficient knowledge of art subjects rendered many of their designs more strange than beautiful, and more noticeably so when they were working on the lines of any given style; but through diligent efforts their designs were steadily improved, and this, in connection with their superior facilities, secured them a large part of the Eastern trade." [29]

Whether or not it was justified, in the late 1870s a reaction set in against poorly produced Eastlake furniture. When Eastlake's fourth revised British edition of *Hints on Household Taste* appeared, in 1878, he felt obliged to disclaim what American tradesmen "are pleased to call 'Eastlake' furniture,

with the production of which I have had nothing whatever to do, and for the taste of which I should be very sorry to be considered responsible." In that same revised edition, which appeared just ten years after *Hints on Household Taste* was first published, Eastlake replaced two of his original furniture illustrations, which were distinctly medieval in feeling, with newer designs. He also noted that the "current of public favor" had shifted, between 1868 and 1878, "from medieval art towards those principles of design which prevailed during the seventeenth and eighteenth centuries, and many a . . . student who began as an earnest advocate for Gothic is now a follower of what is commonly called the 'Queen Anne' school." [30] Beyond these points, the fact that Eastlake did not greatly revise the content of *Hints* after ten years of changing fashion further substantiates the claim that he did not consider his precepts of good taste limited to a single style.

Those who manufactured, sold, and wrote about American furniture in the 1870s and 1880s had to contend with a confusing welter of stylistic labels. The term "Eastlake" was applied to everything from rigidly medieval, massively sober furniture after the manner of Talbert to delicate, Japanese-influenced Herter pieces. "Queen Anne," with its rudimentary allusions to early eighteenth-century design, was considered a fair description of anything having a good amount of spindlework in its construction, while "Elizabethan," "Jacobean," and "early English" were used almost interchangeably to describe squared-off revival pieces with shallow-carved surface ornament. In a grandly ecumenical fashion, the Heywood chair company of Gardner, Massachusetts, advertised "Grecian Queen Anne chairs." The term "art furniture" was used in a generic sense to encompass all the foregoing styles and their hybrids, even when such furniture was machine-produced. By 1883, confusion over names and the styles they presumed to describe led the *Decorator and Furnisher,* a trade journal, to comment: "The confusion in regard to the proper naming of the various styles may also, in part, be charged to the late 'revival' of the 'Queen Anne' and the 'Eastlake' designs, under the meretricious teachings of a few art cranks. The distinctive naming of these two fashions, and the common talk about them for a while, led many people into

[28] Ferris, *Gems of the Centennial,* pp. 132–33.
[29] Gay, "Furniture Industry in America," p. 245.
[30] Eastlake, *Hints,* pp. xxiv, 40.

the habit of naming their furnishings as they would a piece of dress goods." [31]

The stylistic eclecticism implied by this welter of names existed more in the ornamentation of furniture than in its form. It seems probable that all the stylistic variants of art furniture in America owed their existence to the demand for tastefully designed goods initially created by Eastlake's *Hints on Household Taste*. In a broad sense, it seems reasonable to say that all such furniture, whether medieval revival or Queen Anne revival, was also "Eastlake" in the sense of "Eastlake-influenced." Only the most rabid purist, then or now, would insist on using "Eastlake" to describe only the small number of items designed by Eastlake himself or by others in direct imitation of the illustrations in *Hints on Household Taste*.

By the late 1870s, demand for Eastlake-influenced reform furniture had encouraged a large American production of four general types: (1) very expensive furniture turned out to individual order by cabinetmakers and quality decorating companies using custom-production methods; (2) moderately expensive furniture made up from current designs and stocked in small quantities to be sold at retail by the maker-dealer, who would also execute stock designs to special order; (3) well-constructed but moderately priced factory-made furniture produced in the Middle West for wholesale marketing to eastern retail houses; and (4) inexpensive, rather insubstantial factory-made furniture cheaply produced for a working-class market, often sold in the factory's own retail warerooms in eastern cities.

In the 1870s and 1880s, the well-known New York decorating firm of Herter Brothers custom-crafted a number of fine Eastlake-influenced pieces meeting the standards of the first category described above. Though Herter's business was large, most of its furniture was custom-designed and produced by German craftsmen of the old school. An ebonized cherry desk inlaid with light wood, now in the Metropolitan Museum of Art, was made to order for Jay Gould in 1882 for $550 (Fig. 8). Its simplicity and chasteness of design, ebonized finish,

FIG. 8. Herter Brothers, secretary made for Jay Gould. New York City, 1882. Ebonized cherry inlaid with light wood; H. 54". (Metropolitan Museum of Art, gift of Paul Martini.)

and abstracted flower-patterned marquetry all conformed to Eastlake's dicta, although the delicate flat patterning of the marquetry itself was undoubtedly influenced by the rage for orientalia that accompanied the "art movement" in the 1870s and 1880s. The bedroom suite that this desk matched was not, however, unique; around the same time, Herter's made a nearly identical suite for William Carter of Philadelphia.[32]

According to an 1877 newspaper account, John Bond Trevor of Yonkers, New York, commissioned the Philadelphia firm of Daniel Pabst to design a "buffet" to fit a niche in the dining room of Glenview, the mansion he was building on the banks of the Hudson (Fig. 9).[33] Although the galleried top

[31] Heywood Brothers and Company, *Illustrated Catalog* (Gardner, Mass., 1883–84), Metropolitan Museum of Art, New York (hereafter MMA); *Decorator and Furnisher* 2 (Feb. 1883): 137. For examples of factory-made furniture selling in retail warerooms, see Brooklyn Furniture Company, *Trade Catalog* (Brooklyn, 1884), MMA.

[32] Metropolitan Museum of Art, *Nineteenth-Century America: Furniture and Other Decorative Arts* (New York: New York Graphic Society, 1970), no. 210. The Carter furniture is in the collection of the Philadelphia Museum of Art.

[33] *Statesman* (Yonkers, N.Y.), Aug. 3, 1877.

FIG. 9. Sideboard base made for John Trevor Bond, attributed to Daniel Pabst. Philadelphia, 1876. Black walnut; H. 39½″, W. 88″. (Hudson River Museum: Photo, Jon Batkay.)

FIG. 10. Detail of figure 9, panel from door of sideboard. W. 20″. (Hudson River Museum: Photo, Jon Batkay.)

of this sideboard is missing, the base remains in its niche in Glenview, now part of the Hudson River Museum complex. Carved on the door panels of this massive black walnut piece are scenes from Aesop's "Fox and Crane" fable (Fig. 10), taken almost line for line from a design for embroidered curtains by C. Heaton that had previously appeared as an illustration in Eastlake's *Hints on Household Taste* (Fig. 11). Whether or not the Pabst attribution is correct, the Glenview sideboard is obviously the product of a custom cabinetry shop, and at least a part of its design was directly inspired by Eastlake's book.

Kimbel and Cabus, whose exhibit at the 1876 Philadelphia Centennial (Fig. 6) drew favorable reviews, was listed in the New York City directory for that year as "furniture manufacturers" at 7 and 9 East Twentieth Street. Because of the firm's moderate prices and the apparent availability in stock of its furniture, Kimbel and Cabus belongs to the second category of art furniture producers. The designs in the Kimbel and Cabus catalog for 1876–77 were photographed from stock. The only known extant copy of this catalog has penciled price notations above each piece: simple chairs were as low as fifteen dollars, while large case pieces cost three hundred dollars or more. Most of the Kimbel and Cabus pieces produced in 1876 were closely inspired by Eastlake's own designs. They were medieval in feeling, with stained or ebonized finishes,

FIG. 11. C. Heaton, design for embroidered curtains. From Charles L. Eastlake, *Hints on Household Taste*, ed. Charles C. Perkins (Boston: James R. Osgood & Co., 1872), pl. 13. (Hudson River Museum: Photo, Jon Batkay.)

incised ornament that was often gilded, perforated trefoil decorations, and rows of spindling. The ebonized cabinet illustrated here has set-in tiles made by the Minton-Hollins Company of England and brass hinges that are both decorative and functional (Fig. 12). Except for the tiles, which may have been specially ordered to each customer's taste, this piece is identical to the one photographed for the 1876–77 Kimbel and Cabus catalog, where it is accompanied by a price notation of seventy-five dollars.[34] A labeled Kimbel and Cabus library table that was made of ebonized cherry in the 1880s shows Eastlake's influence in its simple squared-off lines, the spindled gallery around its lower shelf, and its modest, shallowly incised geometric and stylized floral decoration (Fig. 13).

While museums hold documented examples of the better Eastlake furniture, it is not easy to find existing mass-produced Eastlake-type pieces that can be attributed to a particular factory. However, illustrations in the furniture trade catalogs issued by many factories in the late 1870s and 1880s amply document factory output in the category of moderately priced furniture made for the wholesale trade. Western factories of this period, influenced by Eastlake's call for honesty in construction, filled their catalogs with glowing reassurances. "All lumber used is of good quality, kiln-dried. We make our patterns and ornamentation of such chaste and elegant forms as suit a cultivated modern taste, and put our work together in the solid, old-fashioned way of our ancestors. There is no sham whatever about any of our work. The cheapest desks are thoroughly and substantially made."[35] This statement, from the A. H. Andrews Company of Chicago, was supported in its 1877 trade catalog by illustrations of chairs and bookcases, simply constructed and ornamented in the manner of East-

[34] Kimbel and Cabus, *Trade Catalog* (New York, 1876). The copy of this album with mounted photographs and handwritten dimensions and prices is held by the Cooper-Hewitt Museum.

[35] A. H. Andrews and Company, *Trade Catalog* (Chicago, 1877), MMA.

FIG. 12. Kimbel and Cabus, cabinet. New York, 1876. Ebonized cherry; H. 50", W. 27". (Hudson River Museum: Photo, Jon Batkay.)

FIG. 13. Kimbel and Cabus, library table. New York, ca. 1880. Ebonized cherry; H. 30½", W. 34". (Hudson River Museum: Photo, Jon Batkay.)

lake's own designs. In general, such pictures in manufacturers' catalogs show that both the midwestern and the eastern factories supplying furniture wholesale to the better retail outlets succeeded in translating Eastlake's tenets of simplicity, honesty, and functionalism into a reasonably priced reality.

Factory-made bedroom furniture was nearly always designed with a chasteness and simplicity well aligned with Eastlake's ideas (Fig. 14). A suite illustrated in the Williamsport Furniture Company's catalog for 1886 could be purchased wholesale for under fifty dollars (Fig. 15). Like other quality factories, the Williamsport Furniture Company was prepared to supply the tasteful woods favored for art furniture: walnut, cherry, oak, "antique" oak, ash, and maple, "finished or in the white." However, the company used applied carved ornament, "artistic, solid and beautiful Mankey Patent Decoration," which Eastlake would have found a laughable contradiction in terms.[36]

Eastlake's proscription of applied carved ornament was seldom observed by American mass-manufacturers, who had decorated their Renaissance revival pieces with trophy swags, acanthus-leaf drawer handles, scrollwork, and medallions purchased from independent "ornament factories," which were equipped with special carving machines to turn out such decorative elements quickly and cheaply. When Renaissance revival furniture gave way to Eastlake-influenced styles, ornament factories began to sell new types of applied decoration in frieze and panel shapes, which afforded mock carved-in-the-wood sincerity to mass-produced furniture. In 1883 the T. B. Rayl Company of Detroit advertised both types of applied ornament for purchase by furniture factories and others.[37]

In the fourth category of furniture manufacture were large factories producing very cheap furniture for retail sale to the working classes, often on

[36] Williamsport Furniture Manufacturing Company, *Trade Catalog* (Williamsport, Pa., 1886), MMA.

[37] T. B. Rayl and Company, *Trade Catalog* (Detroit, 1883), MMA.

FIG. 14. Dresser, probably New York, ca. 1880. Chestnut and ash; H. 79¾″, W. 45″. (Collection of Mr. and Mrs. Paul Piazza: Photo, Jon Batkay.)

FIG. 15. Williamsport Furniture Manufacturing Company, bedroom suite. From Williamsport Furniture Manufacturing Company, *Trade Catalog* (Williamsport, Pa., 1886). (Metropolitan Museum of Art, Whittelsey Fund.)

No. 218.

FIG. 16. Brooklyn Furniture Company, drawing room furniture. From Brooklyn Furniture Company, *Trade Catalog* (Brooklyn, 1884), p. 56. (Metropolitan Museum of Art, gift of Lincoln Kirstein.)

credit. It was here that Eastlake styles were most grossly misinterpreted, if they were attempted at all. Renaissance revival styles retained their popularity with the least affluent portion of the population through the mid-1880s, and many factories met this demand, supplying ill-designed walnut drawing room and bedroom suites on the installment plan to those who had never read Eastlake, Spofford, or Cook. Jordan and Moriarty of Boston, a typical manufacturer of cheap furniture, urged its customers to buy on credit, offering "the very lowest market prices and on such easy terms by weekly payments as . . . means will allow." The drawing room suites illustrated in the 1884 Brooklyn Furniture Company catalog were described as "on the Eastlake order and [as having] many admirers. . . . the frames are of walnut, cherry, or ebony, all at the same price ($55 for seven pieces)"

(Fig. 16).[38] Though the incised and pierced geometric ornamentation on the back crests of these pieces might be considered vaguely "on the Eastlake order," it was probably this sort of furniture that gave all mass-produced Eastlake pieces their reputation for "well-meaning ineptitude." The Brooklyn Furniture Company, which touted itself as "the largest, cheapest, and most reliable furniture maker in the world," had eighteen warerooms in which suites like the one illustrated were sold to a retail audience whose taste-consciousness had yet to be raised.

As shown by the foregoing examples, Eastlake's tenets were indeed applied to "all kinds and conditions of manufacture" in America. By 1883 his in-

[38] Jordan and Moriarty, *Trade Catalog* (Boston, 1885), MMA; Brooklyn Furniture Company, *Trade Catalog*, p. 56.

Fig. 17. J. L. Mott Iron Works, porcelain bathroom fixtures. From advertisement in the *Decorator and Furnisher* 2 (June 1883): 98. (Art and Architecture Division, New York Public Library, Astor, Lenox, and Tilden foundations.)

fluence extended even to the bathroom, as shown by a J. L. Mott Iron Works advertisement in the *Decorator and Furnisher* (Fig. 17). Nor had the magic of the word "Eastlake" escaped the patent furniture designers, whose products made a distinctive contribution to nineteenth-century American furniture manufacture. On September 26, 1876, patent number 182,669 was granted to William Homes of Boston for his "Patent Eastlake Book Case" (Figs. 18, 19), a simple rectilinear form with open adjustable shelves edged with strips of decorated leather following Eastlake's suggestion that "a little leather valance should always be nailed against outer edges [of shelves]. This not only protects the books from dust, but when the leather is scalloped and stamped in gilt patterns, it adds considerably to the general effect." [39] The device for which Homes's patent was obtained, a combination of sliding rails and panels permitting the bookcase to be broken down for portability, may have violated Eastlake's philosophical principle of obviousness of construction, but the mortise and tenon joints holding the uprights in place, as well as the simple, unostentatious design and the geometric

[39] Eastlake, *Hints*, p. 129.

ornament of the case itself, are the essence of Eastlake.

The all-pervasiveness of Eastlake's influence is further demonstrated in the account books kept by the New York cabinetmaker Ernest Hagen from 1880 to 1886. For many years Hagen made furniture for New York's best families, including the Roosevelts; in the 1880s, as sketches in his account books show, he catered to the demand for reform styles at very reasonable cost, turning out items he called "spindle settes," and "Clarence Cooke bookcases," as well as a variety of Eastlake-influenced chairs and tables after Cottier's and Herter's designs. The versatile Hagen would accommodatingly restructure a customer's outmoded Renaissance revival easel, giving it the desired Eastlakean look by removing the curved crest and adding a squared-off, spindled top (Fig. 20), or he would make an entire set of "artistic" chairs to order, as he did in May, 1885, for Louis C. Tiffany. Hagen apparently respected the quality of factory-made bedroom furniture from the Middle West and could certainly not compete with its prices: he recorded the wholesale purchase of suites from Nelson, Matter and Company of Grand Rapids for resale to his customers. He also recorded ordering "5/12 doz. walnut Eastlake

FIG. 18. William Homes, Eastlake portable bookcase. Boston, 1876. Oak; H. 67″, W. 48″. (Hudson River Museum, gift of the Doran family: Photo, Jon Batkay.)

FIG. 19. Labels from back panel of Eastlake portable bookcase shown in figure 18. (Hudson River Museum, gift of the Doran family: Photo, Jon Batkay.)

library chairs" from the New York chair factory of J. W. Mason.[40] A respected and successful cabinetmaker, Hagen would not have easily risked his reputation on factory-produced goods if they did not meet reasonably high standards of design and construction.

In summary, it will be remembered that Eastlake's tenets of honesty in construction, functionalism in design, appropriateness in decoration, and general simplicity and nonostentation grew out of the English design reform movement. His ideas, thoroughly expressed in *Hints on Household Taste,* were echoed in America by the popular press in the mid-1870s and were embodied in some of the British and American furniture exhibits at the Philadelphia Centennial Exhibition of 1876. This widespread dissemination of Eastlake's philos-

ophy created an American demand for furniture of better design and construction, to which manufacturers of every class responded by producing a variety of new artistic styles in general conformity with Eastlake's precepts. It seems unfortunate that the cheapest, most poorly constructed of the factory-made pieces should have inspired a later distaste for all Eastlake-influenced furniture, preventing a fair evaluation of Eastlake's influence, which affected the best classes of furniture manufacture as well as the poorest.

Did a definable "Eastlake" style exist in America between 1870 and 1890? Yes, if that term is used in the reasonably broad sense of "Eastlake-influenced." Much of the art furniture of this period, no matter what its particular stylistic derivation, exhibited certain common design elements favored by Eastlake: rectilinearity of form; use of turned balusters and spindles; chamfered edges; surface decoration of floral and geometric forms either incised, carved

[40] Account Books of Ernest Hagen, 1880–86, entries for May 7, 1885, June 2, 1881, Oct. 4, 1884, New-York Historical Society.

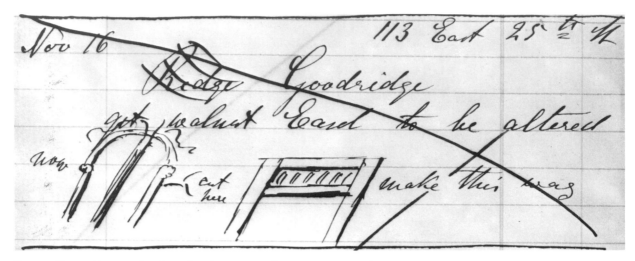

Fig. 20. Ernest Hagen, sketch and written entry from Account Books, 1880–86, entry for Nov. 16, 1886. (New-York Historical Society.)

in shallow relief, or inlaid in marquetry; ebonized or darkly stained finishes enhanced with incised gilding; and general use of decoratively grained light woods.

Did Eastlake-influenced American furniture conform to Eastlake's standards of honesty in construction, function, and ornament? Though Eastlake admired the handcrafted furniture of the Middle Ages, he placed no direct proscriptions on modern methods of construction, including machine work, so long as the end product was a sturdy, honest piece of furniture, without sham or concealment. As trade catalog illustrations and extant pieces of furniture show, much American furniture produced during the 1870s and 1880s was sturdy, obviously constructed, and in general accord with Eastlake's dictates. In the matter of ornament, it seems that only the custom cabinetmakers like Herter Brothers (Fig. 8) and the limited stock producers like Kimbel and Cabus (Figs. 12, 13) hewed directly to Eastlake's philosophy. Perusal of furniture factory trade catalogs and of factory-made Eastlake pieces demonstrates that applied ornament, so deplored by Eastlake, was an almost universal characteristic of factory-made furniture. Although such extraneous ornament was usually well attached (blocked in by molding or recessed into the surface of the wood), it was incontrovertibly in opposition to Eastlake's concept of functionalism. However, applied carved ornament used with restraint was not usually so offensive that it negated all other tasteful qualities of the furniture it adorned.

Finally, was it economically feasible in America to mass-produce and distribute at low cost a type of furniture true to Eastlake's tenets? The simple sturdy furniture that is pictured, described, and priced in trade catalogs of such manufacturers as the Williamsport Furniture Company gives ample evidence that it was. The notion that Eastlake favored only handmade, custom-produced, and therefore expensive, furniture is not borne out by his statements in *Hints on Household Taste.* There was nothing in the processes of machine production inherently incompatible with Eastlake-influenced styles; in fact, American scroll and circular saws, power lathes, mortising machines, and, especially, incising machines were well adapted to turning out the lines and decorative elements of Eastlake furniture.

Apart from his direct influence on American furniture manufacture from 1870 to 1890, Eastlake should be remembered for his role in sharpening the aesthetic discrimination of an entire generation. In awakening the American public to a sense of its interior environment, Eastlake provided an important intellectual bridge between William Morris's English design reform efforts of the 1860s and later efforts of such American arts and crafts movement exponents as Gustav Stickley. *Hints on Household Taste,* together with the dozens of domestic publications it inspired, created a positive climate for an entirely new design aesthetic based on simplicity and utility rather than ostentatious ornamentation.

# Grand Rapids Furniture at the
# Time of the Centennial

*Kenneth L. Ames*

THE CENTENNIAL staged in Philadelphia in 1876 to commemorate America's first hundred years as an independent nation was an event of historic significance. Opened, closed, and punctuated by elaborate ceremonies; hymned, praised, and recorded with pious attention to every detail, the Centennial was America's coming-of-age celebration. Between May 10, 1876, when it opened, and November 10, when it closed, nearly ten million people visited Philadelphia.[1] In hosting this international exhibition, America claimed equal status with the European nations that had given it birth. The evidence of the Centennial vindicated this claim. The entire event was a resounding success. Fairmount Park was adorned with many attractive and innovative buildings; exhibits were numerous and arresting, visitors abundant,

and receipts impressive. The Centennial not only rivaled the previous European exhibitions that had served as models but in many ways surpassed them.[2]

Throughout the second half of the nineteenth century, international exhibitions were of great importance for industry. Reputations were established or lost at these events.[3] Firms that were awarded prizes pointed to them proudly in their advertising as evidence of the high quality of their product.[4] Typically, thousands of exhibitors displayed their goods at an international exhibition, and in Philadelphia in 1876 there were over two thousand exhibits of various aspects of manufacturing from America alone.

Exhibition goods could usually be counted on to demonstrate the best of current taste and to introduce the fashions of the future. Thus visitors to the Centennial saw much with which they were already familiar, as well as a good deal that would become stylish in the next decade.[5] Yet in some ways the

[1] The author gratefully acknowledges financial assistance from Franklin and Marshall College, which aided him significantly in the preparation of this article. For assistance and suggestions, he is also indebted to Robert Boldin, Franklin and Marshall College; Lora B. Cleary, Grand Rapids Public Museum; Wilson G. Duprey and the staff at the New-York Historical Society; Polly Anne Earl, Winterthur Publications; Barbara Sevy, Philadelphia Museum of Art; Lucija Skuja, Grand Rapids Public Library; and Peter Young and Betty Jo Sheirich of the Franklin and Marshall College Library. Accounts of the Centennial and statistics relative to it can be found in J. S. Ingram, *The Centennial Exposition Described and Illustrated* (Philadelphia: Hubbard Bros., 1876); James D. McCabe, *The Illustrated History of the Centennial Exhibition* (Philadelphia: National Publishing Co., 1876); Frank H. Norton, ed., *Frank Leslie's Illustrated Historical Register of the Centennial Exposition 1876* (New York: Frank Leslie, 1876). Much useful information also appears in Dee Brown, *The Year of the Century: 1876* (New York: Charles Scribner's Sons, 1966); John Maass, *The Glorious Enterprise* (Watkins Glen, N.Y.: American Life Foundation, 1973). The attendance figure is for all admissions, including those wth free passes.

[2] See the comparative statistics in Ingram, *Centennial Exposition,* pp. 753–58; McCabe, *Illustrated History of the Centennial,* pp. 885–90. "The Exhibition was a disappointment only in one respect: it failed to make a profit for the holders of pianos, organs, wines, and cigars.

[3] A discussion of international furniture competition appears in Kenneth Ames, "The Battle of the Sideboards," *Winterthur Portfolio 9,* ed. Ian M. G. Quimby (Charlottesville: University Press of Virginia, 1974), pp. 1–27.

[4] Some firms reproduced their medals and awards on the product itself or on its packaging, especially manufacturers of pianos, organs, wines, and cigars.

[5] While the Centennial was important for introducing Japanese forms to Americans, its role in encouraging Queen Anne and colonial revival styles has been disputed. See Clay Lancaster, *The Japanese Influence in America* (New York: W. H. Rawls, 1963); Vincent J. Scully, Jr., *The Shingle Style and the Stick Style* (1955; rev. ed., New Haven: Yale Univer-

exhibits at the Centennial could be misleading. In the furniture courts, for example, there were about 130 American exhibitors. Of these, well over 50 were from Philadelphia, and about 30 were from New York. One might suppose from these figures that the center of the American furniture industry in 1876 was Philadelphia, but that would be erroneous; New York ranked first. Had the exhibition been held there, exhibitors from New York would probably have been the most numerous. But because the Centennial was in Philadelphia, Philadelphians turned out in great numbers, both as exhibitors and visitors.

Also misleading was the preponderance of eastern exhibitors. From the evidence of the Centennial, the manufacture of furniture seemed almost wholly confined to the eastern seaboard. Out of approximately 130 furniture exhibits, only 18 were from the Midwest. Yet the national census figures for 1880 reveal that Chicago, Cincinnati, and Saint Louis ranked second, fourth, and sixth among American cities in the value of the furniture they produced annually.[6] Here again geographic factors undoubtedly played a part. Few manufacturers were willing to undertake the costs and difficulties of sending an exhibit any great distance, particularly if their established markets lay in the South and West rather than the East.

There is irony in the fact that this impression of eastern domination occurred at the very time when eastern centers were first being seriously challenged by the furniture factories of the Midwest. Since early colonial times, fine furniture had been made in the cities of the East, but as settlers began to move west, furniture production did too. By the early years of the nineteenth century, considerable furniture was being produced in the Midwest, and by 1840 Cincinnati was already an important cen-

ter.[7] Immediately after the Civil War, a great expansion took place in the Midwest; population figures began to soar, furniture factories were founded or enlarged, and the relocation of the American furniture industry was accelerated. The New York cabinetmaker, Ernest Hagen, whose memoirs have often been quoted, observed that in the 1870s small eastern cabinetmakers began to feel the effects of competition from big midwestern factories. In a short time the large output and consequent low prices of midwestern factories put the old-time small shop out of business.[8]

One city Hagen cited was Grand Rapids. At the Centennial, three of the five Michigan furniture exhibits were from this city. While they formed but a tiny fraction of the entire furniture display, the three exhibits were significant. They were staged by firms that would become giants of the furniture industry, making the name Grand Rapids synonymous with furniture to millions of Americans. By exhibiting in faraway Philadelphia, these firms demonstrated an ambition and determination that would lead them to commercial success.

These three companies—Berkey and Gay Furniture Company; Phoenix Furniture Company; and Nelson, Matter and Company—are the subjects of this article. Their origins and early development, as well as the furniture they produced, will be examined. It is hoped not only that this discussion will illuminate the accomplishments of these Grand Rapids furniture firms but that it will shed some light as well on the nature of the American furniture industry at the time of the Centennial.

Like so many other midwestern cities, Grand Rapids grew from a tiny settlement to a metropolis in a relatively short time. Although it had been an encampment site for Ottawa Indians from early days, white men first settled along the falls of the Grand River in 1826.[9] By 1836 Grand Rapids num-

sity Press, 1971); Rodris Roth, "The Colonial Revival and 'Centennial Furniture,'" *Art Quarterly* 27, no. 1 (1964): 57–81; Maass, *Glorious Enterprise,* pp. 84–92. The American furniture exhibitors are listed in U.S. Centennial Commission, *International Exhibition 1876 Official Catalog* (Philadelphia: John R. Nagle & Co., 1876), pp. 113–16. Some of the exhibits are described at greater length in Samuel J. Burr, *Memorial of the International Exhibition* (Hartford: L. Stebbins, 1877).

[6] U.S., Department of the Interior, Census Office, *Report on the Manufactures of the United States at the Tenth Census* (Washington, D.C.: Government Printing Office, 1883), p. xxvi.

[7] For an example of western production, see Jairus B. Barnes, "William Tygart, A Western Reserve Cabinetmaker," *Antiques* 101, no. 5 (May 1972): 832–35. An account of furniture manufacturing in 1840 can be found in Charles Cist, *Cincinnati in 1841* (Cincinnati: by the author, 1841). The growth of a decade can be discerned by consulting Charles Cist, *Sketches and Statistics of Cincinnati in 1851* (Cincinnati: W. H. Moore & Co., 1851).

[8] Hagen quoted in Elizabeth A. Ingerman, "Personal Experiences of an Old New York Cabinetmaker," *Antiques* 84, no. 5 (Nov. 1963): 580.

[9] The early history of Grand Rapids is described in Albert Baxter, *History of the City of Grand Rapids, Michigan* (New

bered about five hundred inhabitants. In that year the town's first furniture maker, the near-legendary William ("Deacon") Haldane, came to Grand Rapids from Painesville, Ohio, where he had operated his own cabinet shop. After becoming established in Grand Rapids, Haldane worked there with a variety of partners for most of the rest of the century, turning out goods for the local market. Because he was reputedly the first furniture maker in the city of furniture, and because, living to a ripe old age, he represented a living link with pioneer days, Haldane figures prominently in most histories of the origins of Grand Rapids furniture. In some accounts his name is surrounded by an aura of reverence.[10]

Deacon Haldane was soon followed by a number of cabinetmakers who were undoubtedly attracted by the many advantages of Grand Rapids. There were vast supplies of wood readily at hand. A large river provided easy means of transportation both for logs and for the finished product, and the river rapids provided ready power for machinery. Steamboats had begun to navigate the river in 1837, and by 1858 the railroad had come to Grand Rapids. Within a few years the city was served by four lines.[11] Based on these tangible factors, the lure of financial success drew men to Grand Rapids.

At the middle of the century, half a continent lay waiting to be exploited, and a swelling nation of ever more prosperous citizens was eager to be supplied with consumer goods. Images of a national market must have danced in the minds of the early industrialists of Grand Rapids, for the city itself certainly provided little outlet for goods produced

there. In 1860, numbering 8,085 inhabitants, Grand Rapids could only with difficulty have supported the nine firms then engaged in the production of furniture. In that year these firms paid 53 workers wages totaling $14,808 to turn out a product valued at $32,255, a good deal of which was marketed in other cities.[12]

In the next decade, from 1860 to 1870, an ever increasing portion of the industry's product was sold outside the city. Although the Civil War dominated half of the period, the furniture factories expanded at a swift rate. While the number of firms in 1870 had declined to eight, all other figures associated with the furniture trade had grown impressively. The number of employees had increased to 281, wages had swelled to $131,500, and the product was valued at $348,900. Yet during this same time the population of the city had only doubled, to about 16,500.[13]

The decade of the 1870s witnessed an even more dramatic expansion of the furniture industry in Grand Rapids (see table 1). This was the decade in which the city made its mark; this was the decade that saw it rise to national significance. In 1870 the factories of Grand Rapids were responsible for only 0.5 percent of total furniture production in the country. By 1880 this figure had risen to 2.5 percent, and Grand Rapids ranked seventh nationally in the production of furniture.[14] Although in later years the city grew to greater national prominence and assumed a larger percentage of the trade, it was in the 1870s that those giant firms, so crucial to the development of Grand Rapids and to the history of the nineteenth-century American furniture industry, expanded and incorporated. After 1880 the rate of expansion began to decline, although Grand Rapids remained an active furniture center well into the twentieth century.

The evolution of the three great Grand Rapids firms hinged on the activities of a handful of men. Although records are scattered and much of the early history of furniture in the city is based on the later recollections of old men, it is generally agreed

York: Munsell & Co., 1891); Dwight Goss, *History of Grand Rapids and Its Industries* (Chicago: C. F. Cooper & Co., 1906).

[10] The basic works on the early history of furniture manufacture in Grand Rapids are, in chronological order: Baxter, *History of the City of Grand Rapids* (1891); Goss, *History of Grand Rapids* (1906); William Widdicomb, "The Early History of the Furniture Industry in Grand Rapids," *Publications of the Historical Society of Grand Rapids* 1, no. 5, pt. 5 (1909): 63–76; Frank Edward Ransom, *The City Built on Wood, A History of the Furniture Industry in Grand Rapids, Michigan, 1850–1950* (Ann Arbor: Edwards Bros., 1955); James Stanford Bradshaw, "Grand Rapids Furniture Beginnings," *Michigan History* 52, no. 4 (Winter 1968): 279–98. Some of Goss's account is a verbatim repetition of Baxter's.

[11] The Lake Shore and Michigan Southern, the Detroit and Milwaukee, the Grand Rapids and Indiana, and the Michigan Central. See Ransom, *City Built on Wood*, p. 14.

[12] Five sets of census figures (1860–1900) are conveniently summarized in Ransom, *City Built on Wood*, p. 13.

[13] Population statistics are listed in Grand Rapids Board of Trade, *Grand Rapids as It Is* (Grand Rapids: Eaton, Lyon & Allen Printing Co., 1888), p. 6.

[14] For useful tables, see Bradshaw, "Furniture Beginnings," p. 298; U.S., Department of the Interior, Census Office, *Report on Manufactures*, pp. xvii, xxvi, 40–41, 403–4.

that the beginnings of the important Berkey and Gay firm can be traced to 1859. In that year Julius Berkey, in partnership with a James Eggleston, began production of sash, doors, and blinds. Later that year or early in 1860, he changed partners and

ship, taking Elias Matter into the business. The outlook for this new venture was not particularly promising. Berkey had already failed once. At the time of the new partnership he reputedly had only five dollars in cash and a few hundred dollars in

Table 1. Grand Rapids Furniture Industry, 1860–80

| Year | Number of furniture firms | Number of employees | Wages | Capital invested | Costs for materials | Value of product |
|---|---|---|---|---|---|---|
| 1860 | 9 | 53 | $ 14,808 | $ — | $ — | $ 32,255 |
| 1870 | 8 | 281 | 131,500 (90,735) | 329,500 (227,355) | 117,650 (81,178) | 348,900 (240,741) |
| 1880 | 15 | 2,279 | 720,000 (669,600) | 1,636,000 (1,521,480) | 908,000 (844,440) | 2,016,000 (1,874,880) |

NOTE: The figures within parentheses represent the current dollar values adjusted to 1860 constant dollars using the Warren and Pearson wholesale price index. U.S., Department of Commerce, *Historical Statistics of the United States: Colonial Times to 1957* (Washington, D.C.: Government Printing Office, 1961), p. 115.

with Alphonso Ham (or Hamm) began to make furniture for the Chicago market. They worked in a portion of the second floor of a factory just erected on Mill Street by Julius's brother William for the manufacture of sash, doors, and blinds. Berkey and Ham installed a lathe and the first shaper (a machine for cutting irregular shapes from flat pieces of wood) in Grand Rapids[15] and fabricated a variety of pieces of furniture. One of their more successful products was a small walnut table, inexpensive to make yet "showy" enough to pass for a more expensive piece, which later came to be known as a "Berkey table," at least in Grand Rapids.[16]

The Berkey and Ham association soon failed and was dissolved. In April 1861, Julius Berkey was again producing furniture on the second floor of brother William's factory. A year and a half later, in November 1862, he formed his third partner-

machinery and materials. Elias Matter brought his tool chest and its contents, valued at six dollars, to the association.

Such unimpressive assets were typical of the early days of many of the great furniture concerns of the last century. Yet with their modest assets, Berkey and Matter prospered from the start. They directed their attention to the manufacture of furniture specifically for the wholesale trade in Milwaukee and Chicago, where it was well received. By October 5, 1863, William Berkey was lured into taking a half interest in the firm, which was renamed Berkey Brothers and Company.[17]

William had apparently accumulated enough in profits from his factory to invest substantial capital in his brother's venture. After his entry into the firm, its capital was listed at over $17,000, including real estate, machinery, and materials. In effect, William's partnership may have represented a partial takeover of Julius's firm, giving the latter more capital and William a large share of the profits. This suggestion is supported by the fact that Berkey Brothers manufactured sash, blinds, and doors, William's staples, and furniture, Julius's specialty.

The partnership was maintained through the Civil War until 1866, when George W. Gay purchased half of William Berkey's interest (or one-

[15] Baxter, *History of the City of Grand Rapids,* p. 465; Widdicomb, "Early History," p. 65; Ransom, *City Built on Wood,* pp. 11–12, 88; Bradshaw, "Furniture Beginnings," p. 290. The purpose here is not to unravel the intricacies of the founding and growth of these firms but rather to sketch the broad outlines as they are generally seen. Most Grand Rapids furniture history takes the form of personal reminiscence; for this reason, there is considerable disagreement among the various sources about the details of these firms' histories.

[16] The exact appearance of the "Berkey table" is not clear. Widdicomb, "Early History," p. 66, described it as "a walnut table . . . a little, inexpensive affair." Bradshaw, "Furniture Beginnings," p. 290, quoting an earlier source, mentioned "small walnut scallop stands with tops."

[17] Widdicomb, "Early History," p. 65; Baxter, *History of the City of Grand Rapids,* p. 465; Bradshaw, "Furniture Beginnings," p. 293.

quarter of the firm). This occasioned another name change, from Berkey Brothers and Company to Berkey Brothers and Gay. On February 28, 1870, Elias Matter retired from this association, but not from the manufacture of furniture, and in January 1873 William Berkey withdrew.[18] The Berkey and Gay Furniture Company was incorporated in August 1873—ironically, a year of widespread financial panic. Julius Berkey was named president and George W. Gay treasurer.

As early as 1871, Berkey Brothers and Gay found its quarters in William's factory too restrictive. In December of that year construction started on a five-story brick building measuring 90 by 100 feet. It was to be heated by steam and to be served by two steam-powered elevators. Machinery was to be confined to the older building; the top three floors of the new building were to be used for assembling, upholstering, and finishing furniture, while the two lower floors were to serve as warehouse and showrooms. It was estimated that in addition to the 150 people normally employed, 50 new employees would be required to staff this building. A report of 1874 noted that Berkey and Gay constantly employed "as many as 200 hands," and in 1873 the company claimed to have shipped "about $300,000 worth of furniture to New York, Pennsylvania, Ohio, Indiana, Illinois, Wisconsin, Iowa, Minnesota, Nebraska, Missouri, and Colorado."[19]

The business prospered, and in 1874 Berkey and Gay found it necessary to erect yet another building to house the rapidly expanding establishment. This new structure, hailed as "one of the most prominent features of the city," was considerably larger than either of the earlier two. It had a front of 74 feet on Canal Street, 220 feet on Hastings Street, and 100 feet on Kent Street. Including the basement, the building housed six stories. A contemporary account described it as "substantially constructed, with solid walls of handsome, white brick, and a gravel and tin roof. It is heated throughout by steam and lighted by gas. The build-

FIG. 1. Berkey and Gay Furniture Company, furniture showrooms. From an undated stereograph card. (Grand Rapids Public Library.)

ing is also supplied with water from the city water works, the reservoir being near at hand on the hill overlooking the city, and has standpipes and hose connections on every floor, giving first-class protection against fire."[20]

The interior of the building was subdivided to suit the varied requirements of the furniture business. On the Canal Street front the first floor and basement accommodated a salesroom for general stock and also the company's offices, which were elegantly furnished in black walnut. On the second floor a space approximately 62 by 220 feet was used as a salesroom, "where a full line of samples" was kept. An undated stereograph of this salesroom survives, revealing an extensive and imposing array of furniture (Fig. 1). The three floors above this salesroom served primarily as storerooms for finished as well as unfinished goods.

On the Kent Street end of the building, 100 feet in length, a fire wall divided the building into two parts: the main section, just described, and a smaller section to the south. The basement of this smaller section contained boilers, sealed off by iron doors as protection against fire, for heating the entire complex. The ground floor was used as a packing and trimming room, the next floor as an up-

[18] Baxter, *History of the City of Grand Rapids,* p. 465; Widdicomb, "Early History," p. 67; Bradshaw, "Furniture Beginnings," p. 293.

[19] Ransom, *City Built on Wood,* p. 15; J. D. Dillenback, *Grand Rapids in 1874, Sketches of the Trade, Manufactures and Progress of the City* (Grand Rapids: by the author, 1875), p. 49. The states are probably listed in order of the value of their sales. The placing of New York at the head of the list lends support to Hagen's claim that Grand Rapids had invaded the New York market by the early 1870s.

[20] Dillenback, *Grand Rapids in 1874,* p. 17.

holstering and sewing room, and the three rooms above as finishing rooms.[21]

There were two noteworthy inclusions in the Berkey and Gay factory. The first was the elevator. Manual elevators, relying on ropes, pulleys, and hand cranks, date back to antiquity, but it remained for the nineteenth century to invent the mechanical elevator, an invention in which America played an important role.[22] At New York's Crystal Palace Exhibition of 1853, Elisha Graves Otis displayed "the world's first safe elevator," a patented device he had produced by adding a safety catch to an ordinary hoisting platform. Elevators thereafter gradually became more common, especially from the early 1870s onward.[23]

Berkey and Gay's 1871 building had contained two steam-powered elevators, and its new building also contained two, each placed with an eye to reducing labor, time, and costs. In the southern section of the building there was a power elevator at the end of a covered alley, allowing goods to be loaded and unloaded in a sheltered area and then taken to the appropriate floor. Another elevator, described as "a large power elevator," was near the salesrooms on the Canal Street side. It was intended to carry both passengers and freight. There was also a hand elevator in the center of the building for conveying goods from floor to floor.

The second important inclusion in this factory was a provision for taking photographs. In the center of the building, and raised one story above the roof, was a small extension or annex. Approximately 20 by 50 feet, this structure was designed as a studio in which photographs of furniture could be made. Pieces of furniture could be brought up to the studio by elevator, photographed, and then returned to the showrooms. From the glass collodion negatives, paper prints could be made, in great numbers if desired. These were gathered in books and taken around the country by the company's salesmen.[24]

It is not certain when photography was first used for commercial purposes. Multiple salesmen's catalogs with duplicate prints were impractical until the 1850s, when the glass negative process replaced the more fragile and expensive daguerreotype. The earliest photographic furniture catalog known is that of Ribaillier aîné et Paul Mazaroz of Paris, dated 1855–56 and preserved in the Bibliothèque Nationale. Furniture makers in America used photographs as early as 1864, when the Boston cabinet-maker Sidney Squires offered to show Trevor Park photographs of his products. The earliest surviving American catalogs are undated, although the style of furniture in them suggests a date between 1860 and 1865. How early Berkey and Gay used photography is not clear. Elias Matter is credited with introducing the use of photographic salesmen's catalogs to Grand Rapids in 1862. "Matter is said to have obtained the idea from a salesman for baby carriages whom he met in Jackson, Michigan." [25] In any case, the rather elaborate provisions made for photography in Berkey and Gay's 1874 building reveal that by that time the firm was fully aware of its advantages. Use of the elevator and photography are indications of Berkey and Gay's willingness to utilize technological advances in the business, a willingness that must have contributed to the firm's commercial success.

While Berkey and Gay was passing through its formative phase, a second major firm was evolving in Grand Rapids; and Julius Berkey's brother William played the dominant role in this evolution. William A. Berkey, born in 1823 on a farm in Perry County, Ohio, spent his early years working as a teacher during the school year and as a carpenter during the summer. In 1848 he got married and set himself up as a manufacturer of sash and doors in Tiffin, Ohio. He continued this trade for seven years, at which time he sold out and moved to Grand Rapids with the intention of resuming

[21] Dillenback, *Grand Rapids in 1874*, pp. 17–18.

[22] For early examples of the use of power elevators, see Henry-Russell Hitchcock, *Architecture, Nineteenth and Twentieth Centuries* (Harmondsworth: Penguin Books, 1968), p. 85; Carl W. Condit, *American Building* (Chicago: University of Chicago Press, 1968), p. 84.

[23] Siegfried Giedion, *Space, Time, and Architecture* (Cambridge: Harvard University Press, 1962), p. 207; Hitchcock, *Architecture*, p. 239.

[24] Dillenback, *Grand Rapids in 1874*, p. 17. Examples of Grand Rapids furniture catalogs with photographs are preserved in the Grand Rapids Public Library.

[25] Ribailler et Mazaroz catalog, 1855–56, Print Room, Bibliothèque Nationale, Hd 108 and Hd 108a (folio); Squires to Park, Park-McCullough Mansion files, North Bennington, Vt. An early undated, unattributed volume of furniture prints is in the Winterthur Museum Libraries, MS 2944, 68 x 84, and there is one in the Landauer Collection of the New-York Historical Society. Fragments of a third are at the Pennsylvania Farm Museum, Lancaster. Beaumont Newhall, *The History of Photography* (New York: Museum of Modern Art, 1964), pp. 23–26, 28, 31–40; Ransom, *City Built on Wood*, p. 19; Baxter, *History of the City of Grand Rapids*, p. 462.

the same business there. In 1859 he built the factory on Mill Street, part of which his brother occupied for a time. His association with Julius after 1863 has already been discussed. In 1868 he was appointed assignee for the bankrupt firm of Atkins, Soule, and Company. His efforts to save the firm ended when a fire caused heavy damage to the plant. In 1870 the remains of Atkins and Soule were purchased by a group with William Berkey as its president. Since it arose from the ashes of Atkins and Soule, this new firm took the name of Phoenix Furniture Company. Having extensive capital, William Berkey and his associates succeeded from the start. In the autumn of 1872 ground was broken for a new factory, which cost $160,000. Of this sum, $30,000 went for land (some six to eight acres), $75,000 for the main building, $35,000 for a 200-horsepower steam engine and machinery, $15,000 for the sawmill, and $5,000 for the dry kiln.[26] The inclusion of the last two items was not typical in previous furniture factories and is an indication of an early attempt at vertical integration of productive processes in the furniture industry.

The main block of the new brick factory building was four stories high and measured 74 by 200 feet. Boiler and engine rooms were attached. Noteworthy were the elaborate precautions taken to avoid damage by fire, particularly significant considering the origins of this company as well as the frequency with which fire destroyed property in the nineteenth century. William Berkey claimed that the building was, in fact, fireproof. Spaces were carefully insulated so that fire could not spread through the building. A fire wall ran through the center of the building, pierced only by heavy iron doors, and the stairway and elevator compartments were similarly surrounded by heavy brick walls. Although the building used wooden structural members on the interior, the beams were apparently sheathed in sheet iron and the posts coated with stucco. Unfortunately, it was soon discovered that the latter had an alarming propensity for rapid decay and had to be replaced.

It was conventional to convey power from one story to the next by belts passing through openings in the floors, but in Phoenix's factory there were no belt openings. Instead, the main belt ran to shafts

extending outside the exterior walls. This unusual arrangement was the result of William's commitment to his own system of fireproofing, which had its financial advantages as well. The extra expenditures for this unorthodox fireproofing, Berkey believed, would be more than offset by the savings in fire insurance. Because the spaces within the building were so well insulated, only a major fire could burn more than half a floor or so at one time. Thus it was necessary to carry only $10,000 to $30,000 worth of fire insurance as opposed to the $150,000 or more that was normal practice. "Since yearly premiums for such insurance averaged between seven and ten per cent of the total amount, savings up to $14,000 per year could be realized." [27]

As noted before, the Phoenix Furniture Company was nearly self-contained, since it maintained its own sawmill as well as its own kilns for seasoning wood. As one writer stated, "The company take their own timber from the stump, convey it to the premises by water or rail, and saw and season it upon their own grounds." [28] The advantages of this arrangement are obvious. Not only was the sawed and seasoned lumber convenient to the factory and essentially guaranteed to be of good quality and in regular supply, but it was obtained without depending on, or paying, any middlemen along the way. In short this integration of raw material supply and manufacturing processes was recommended by both efficiency and economy.

The actual layout of the sawmill, kiln, and factory showed considerable ingenuity. Tracks were run in to the factory grounds to deliver timber to the sawmill, where it was sawed into boards, which then went to the kilns. The patented kilns were immense affairs, 36 by 80 feet, capable of holding 90,000 feet of lumber stacked on cars. The air inside them was heated by pipes filled with waste steam from the steam engine. Using waste steam for kilns was an extension of the practice of using excess steam to heat factories. From the kiln, further tracks led to the factory. To eliminate the need for motive power, all these tracks, from mill to kiln to factory, ran downhill.[29]

This complex of buildings was only the first stage of a multistaged building campaign. In 1874

[26] Baxter, *History of the City of Grand Rapids*, pp. 467–68; Goss, *History of Grand Rapids*, pp. 1064–65; Ransom, *City Built on Wood*, p. 32.

[27] Dillenback, *Grand Rapids in 1874*, pp. 45–46; Ransom, *City Built on Wood*, p. 32.
[28] Dillenback, *Grand Rapids in 1874*, p. 46.
[29] Ransom, *City Built on Wood*, p. 32.

it was noted that when the main building was completed "in accordance with the original design," it would be 420 feet long, or more than twice the length of the block erected in 1872–73. Additions to the Phoenix plant were made in 1875, 1880, and 1883.[30]

The third of the great Grand Rapids firms, Nelson, Matter and Company, emerged at about the same time as Berkey and Gay and Phoenix. It also evolved slowly from a smaller firm and survived the shifting affiliations of several partners. In 1854 E. W. Winchester came to Grand Rapids from Keene, New Hampshire, and entered into partnership with the venerable William Haldane. This partnership was of brief duration, and in 1855 Winchester formed a partnership with his brother, S. A. Winchester, calling the firm Winchester Brothers. On September 15, 1857, Charles C. Comstock, one of the city's largest and most powerful lumber and wood entrepreneurs, purchased the firm and ran it himself until 1863. In that year he disposed of half his interest to James M. and Ezra T. Nelson, and the name of the firm was changed to Comstock, Nelson and Company. In August 1865, two foremen, Manly G. Colson and James A. Pugh, each purchased an eighth interest in the firm from Comstock, who gave his remaining quarter interest to his son, Tileston A. Comstock. As a result of this shuffling, the name of the firm was changed to Nelson, Comstock and Company. On April 16, 1870, Elias Matter, who had just left Berkey Brothers, purchased the younger Comstock's interest, and the firm then became Nelson, Matter and Company. It continued to bear this name through most of its history, although there were subsequent shifts in the inner structure of the company.[31]

In 1868 the company erected a building on Canal Street for retail showrooms and offices. It contained three stories and a basement and measured 54 by 80 feet. To this relatively small building, a large factory, a boiler, an engine house furnished with a 150-horsepower steam engine, and dry kilns were added in 1873. The main building of this complex was brick and measured 70 by 160 feet. Including the basement, it was five stories high. Like the other factories described, this one was heated by steam and lighted by gas. Goods were moved from floor to floor by a huge steam elevator situated in the center of the building.

One of the most distinctive features of the Nelson, Matter and Company factory was the use of iron tracks to move large amounts of heavy goods. Tracks running to the center of the building were laid through each of the three entrances to the main floor of the factory, so cars loaded with lumber or furniture could be run in and out of the building. The cars could be shifted from one track to another with relative ease on a turntable in the middle of the building, where the tracks converged. Near the convergence of the tracks was the steam elevator, on which materials could easily be brought to or from the various floors of the factory.[32] Thus this factory, like the other large, well-designed industrial plants of Grand Rapids, was well equipped to compete with the older establishments of the East.

Berkey and Gay, Phoenix, and Nelson, Matter and Company all survived the panic of 1873—possibly because they carried on a large enough trade with Canada to see them through—and went on to rapid expansion in the years following. By 1874 Grand Rapids furniture companies were employing 720 workers, their capital was estimated at about $1,150,000, and they were turning out furniture valued at $1,350,000, or nearly $1,000,000 more than in 1870.[33] Credit for most of this growth belongs to the three giant firms.

For these companies, and for Grand Rapids, the Centennial of 1876 was a major victory. Although anglophile critics like Charles Wyllys Elliott held the style of Grand Rapids furniture in contempt, and few compilers of books about the Centennial thought enough of the entries of these firms to include illustrations of them or even mention them, the opinion of the judges and of the general public was quite different. All three of the major Grand Rapids furniture companies received awards at the Centennial.[34] They had obtained considerable

[30] Dillenback, *Grand Rapids in 1874*, p. 46; Baxter, *History of the City of Grand Rapids*, p. 467.

[31] Baxter, *History of the City of Grand Rapids*, p. 462; Bradshaw, "Furniture Beginnings," pp. 287–89.

[32] Baxter, *History of the City of Grand Rapids*, p. 463; Dillenback, *Grand Rapids in 1874*, p. 42. This is an interesting use of the turntable, a device more usually associated with the railroad.

[33] Ransom, *City Built on Wood*, p. 17; Dillenback, *Grand Rapids in 1874*, p. 8.

[34] Elliott, "Household Art, I—The Dining Room," *Art Journal*, n.s. 1 (1875): 295–300. Burr, *Memorial of the Inter-*

practice in setting up exhibitions through their participation in local trade fairs, and it is evident that their entries at Philadelphia were the result of careful and lengthy planning.

Perhaps the most spectacular exhibit was by Nelson, Matter and Company. The firm received an award for an elaborate bedstead and dressing case. Although the prize-winning furniture has been lost, its appearance is known from descriptions and illustrations (Fig. 2). The two pieces, exceptionally ornate variations of contemporary chamber pieces, bore little resemblance to the normal products of the firm. Both pieces were richly encrusted with standing allegorical figures, in a Gothic-Renaissance style vaguely reminiscent of the sixteenth century yet unmistakably of the nineteenth. The largest figure, in a splendid Gothic niche, depicting George Washington, Father of His Country, was wholly appropriate for a Centennial commemorative. Both bed and dressing case were crowned with a carved likeness of the American bald eagle.[35]

The chamber furniture exhibited by Berkey and Gay is thought to be that now in the Grand Rapids Public Museum (Fig. 3).[36] These superb pieces, not wholly unlike some of the most costly in general production, are still distinctive enough to make it clear that the old tradition of *tour de force* production for exhibitions was well understood and appreciated in Grand Rapids. As great displays of design and technical virtuosity, these pieces were intended to dazzle spectators and attract prospective customers.[37] They apparently succeeded, for at about the time of the Centennial, Grand Rapids firms opened showrooms in New York to market their goods on the East Coast. It may have been the displays at the Centennial and the bronze medals they were awarded that attracted foreign customers to Grand Rapids. Shipments went to Europe and South America, and one source noted that in 1879 "a leading hotel in Dundee, Scotland," was reportedly "furnished throughout with the products of Grand Rapids factories."[38]

At about the same time, the importance of Grand Rapids to the American furniture trade was further acknowledged when buyers began to visit the city regularly. The semiannual furniture market, which later grew to immense proportions, drawing hundreds of furniture buyers from all over the country, may have had its modest beginnings in 1878. The newspaper listings of out-of-town arrivals indicate that in late December of that year, eleven buyers from Chicago, Philadelphia, Boston, Milwaukee, and Toledo were present. As early as 1881, the furniture market had become so important that firms from other localities found it profitable to show there.[39]

The 1880 census figures for furniture manufacturing in Grand Rapids have already been given. However, the individual schedules provide a more exact impression of the relative size and importance of the three firms. Berkey and Gay listed its capital

---

*national Exhibition*, p. 305, mentioned and expressed admiration for Grand Rapids products: Nelson, Matter and Company was "commended for good workmanship and finish, and choice selection of materials"; Berkey and Gay "for good work, carefully selected material, and superiority in the details of manufacture"; and Phoenix "for good workmanship and finish, and for adaptation to the demands of the market for which it is manufactured; a fine exhibit." See U.S. Centennial Commission, *International Exhibition, 1876: Reports and Awards*, 6 vols. (Washington, D.C.: Government Printing Office, 1880), 4:733–34. The validity of these awards is discussed in Maass, *Glorious Enterprise*, pp. 114–17.

[35] Nelson, Matter and Company, Centennial award certificate, Grand Rapids Public Library.

[36] This furniture is well known. An illustration of it was first published in Raymond F. and Marguerite Yates, *A Guide to Victorian Antiques* (New York: Harper Bros., 1949), and subsequently appeared in *Victoriana: An Exhibition of the Arts of the Victorian Era in America* (Brooklyn: Brooklyn Museum, 1960); Helen Comstock, *American Furniture* (New York: Viking Press, 1962); Joseph Aronson, *The Encyclopedia of Furniture* (3d ed.; New York: Crown, 1965); Joseph T. Butler, *American Antiques, 1800–1900* (New York: Odyssey Press, 1965); Celia Jackson Otto, *American Furniture of the Nineteenth Century* (New York: Viking Press, 1965); Marshall B. Davidson, ed., *The American Heritage History of Antiques: From the Civil War to World War I* (New York: American Heritage, 1969). The inclusion of the furniture in a recent textbook on American art, Daniel M. Mendelowitz, *A History of American Art* (New York: Holt, Rinehart & Winston, 1970), will make it even more widely known.

[37] They continue to dazzle viewers today. The comments of visitors to the Grand Rapids Public Museum testify to the powerful presence of this furniture. Children are usually sure it belonged to royalty.

[38] Ransom, *City Built on Wood*, pp. 17, 19; Victor S. Clark, *History of Manufactures in the United States*, 2 vols. (New York: McGraw-Hill, 1929), 2:484. According to New York City directories, Phoenix Furniture Company had a New York store from 1877 until after 1925. Nelson, Matter and Company was listed from 1879 to 1890, but Berkey and Gay was listed only for 1878–79.

[39] Ransom, *City Built on Wood*, pp. 19, 22–23.

Fɪɢ. 2. Nelson, Matter and Company, exhibit for the Philadelphia Centennial of 1876. From a trade card. (New-York Historical Society, Landauer Collection.)

as $320,000, its average number of employees as 400, and the value of its product as $525,000. Phoenix claimed to have about $305,000 in capital, to normally employ 520 workers ten hours per day, and to have produced furniture valued at about $514,000. Nelson and Matter held over $400,000 in capital, employed about 380 hands on the average, and valued its product at $315,000.[40] Among them, these three firms were responsible for approximately 60 percent of the city's production of furniture.

The furniture manufacturers of Grand Rapids did not rise to prominence without a keen understanding of American taste. Writers attempting to account for the success of Grand Rapids have correctly emphasized the great attention paid to design. As a writer of 1887 noted: "Each establishment maintains its own staff of designers and they are busy the whole year round planning articles of

furniture as comfortable, unique, and beautiful as the art of man can compass."[41] Although information about designers in the 1870s is scarce, the design of products must have been an important concern.

Grand Rapids furniture had to be designed to meet two special requirements. First, it had to resemble furniture with which people were familiar yet be in some way novel or unique. Second, it had to look expensive but involve only modest expense to the manufacturer. In order to meet the last requirement, the designers needed a thorough knowledge of the capabilities of the machines in the factory. The more work done by machine, the less expensive the piece. If an elaborate piece of furniture could be made primarily by machine, with a minimum of hand labor, it could be offered to the public at a modest price. The production in volume made possible by mechanization also worked to lower prices.

The Philadelphia furniture entrepreneur George Henkels had complained in 1861 about industrial-

[40] Individual Manufacturing Schedules for the Tenth National Census (1880), normally housed in the Michigan State Archives, but now in Washington, D.C. at the National Archives for microfilming. The author is indebted to Polly Anne Earl for pointing out the availability and usefulness of these documents.

[41] *Free Press* (Detroit), Nov. 1887, quoted in Baxter, *History of the City of Grand Rapids,* p. 462.

FIG. 3. Berkey and Gay Furniture Company, chamber suite exhibited at the Philadelphia Centennial of 1876. (Grand Rapids Public Museum: Photo, Brooklyn Museum.)

ists who turned out fancy but cheap articles by machine. Their furniture caught the eye of the public and sold readily because it was so much cheaper than "fine" furniture made largely by hand in the old manner. While Henkels criticized this approach,[42] there was little he could do to prevent the spread of machine-made furniture, for its cheapness satisfied both the producer and the consumer.

This description of the Grand Rapids approach

to furniture design is not meant as condemnation, although some critics have dismissed Grand Rapids design efforts. By simple means, and through the imaginative use of a limited number of devices, some fine designs were created. Restricted means have never been a detriment to art, and the imaginative artist has worked wonders within the limits of the rigid form of the sonnet, the number of ways a brick can be laid, and the requirements of the lithographer's stone. Even if we adopt the most cynical posture and assume that the primary goal of Grand Rapids manufacturers was to make money, the city's furniture firms still provided furniture for hundreds of thousands of people. For this furniture to sell in a highly competitive market, it had to be appealing. To be appealing, it had to be carefully designed to meet the tastes of the

[42] Samuel Sloan, *Homestead Architecture* (Philadelphia: J. B. Lippincott & Co., 1861), p. 328. Elsewhere it has been suggested that Henkels wrote the furniture sections in Sloan's book. See Kenneth Ames, "George Henkels, Nineteenth-Century Philadelphia Cabinetmaker," *Antiques* 104, no. 4 (Oct. 1973): 641–50.

time. And because it was well designed, it satisfied then and is still satisfying today.

During the 1870s, the dominant mode for furniture was an eclectic style loosely called Renaissance. George W. Gay, who must have been well informed on the matter, stated that when machinery became widely used in furniture shops "manufacturers looked for a fashion in which they could use their facilities to the best advantage, and at the same time retain the attractiveness of their earlier work. This they found in the Renaissance, which for a number of years superseded all other styles in the best class of furniture." [43]

The Renaissance style, largely French-inspired, represents what Carroll L. V. Meeks has called synthetic eclecticism, an imaginative selecting and combining of elements from a variety of historical sources. The eclecticism of Renaissance styles is one phase of the larger movement of picturesque eclecticism, which dominated the entire century.[44] Thus in addition to being an eclectic mixture of details—Renaissance, baroque, classical, and wholly invented—the Renaissance style of the nineteenth century is also picturesque. Furniture in this style has agitated outlines, rich and lively surfaces, and generally aggressive designs. At times such furniture is deliberately awkward, clumsy, or ungainly. To eyes nourished on the designs of the eighteenth century, it often appears top-heavy and malproportioned. To nineteenth-century eyes these alternatives to traditional rules for organization or proportion provided visual delights of a highly enjoyable sort.

Because of the interconnections of the partners in the three major Grand Rapids firms, it is not surprising that the products of the firms are quite similar. No clear way of distinguishing their furniture has appeared. This situation is not peculiar to Grand Rapids. Factory-made furniture throughout the country had strong similarities. Indeed, it often seems that only those forms that were patented or those that involved some idiosyncrasy or mannerism can be easily distinguished.[45] Difficulties in attributing furniture are not new, and furniture made in the 1870s seems as hard to attribute as that

produced earlier. A possible reason for difficulties of attribution in the 1870s is the widespread use of the same woodworking machines by factories scattered around the country. It may well turn out that there are Morellian characteristics even in the use of these machines, but they are likely to be subtle.[46]

Because so few authenticated pieces of Grand Rapids furniture of the 1870s survive, the best way to discuss the products of these firms is to rely on their illustrated trade cards and, better still, the photographic albums used by salesmen. The Grand Rapids Public Library preserves a large album for the Phoenix Furniture Company, dated 1878, and several albums for Nelson, Matter and Company, the earliest of which may be from 1876.[47] By a trick of fate, Berkey and Gay, allegedly a pioneer in the use of photography, is not represented. Of the catalogs that do survive, the richest furniture illustrations of the 1870s are in the Phoenix catalog.

In examining the photographs of this furniture, two observations can be made. First, the designers created bold and striking designs with relatively limited means. Second, within the broad stylistic framework of the Renaissance, there is an immense and impressive variety of designs. Both of these points are clearly illustrated in a useful photograph of eight footrests from Phoenix's catalog (Fig. 4). All are seen from the same view; all are about the same size; all are shown in the same state—the wood unvarnished and the seats not yet upholstered. Yet despite this conformity the designs of the pieces are considerably varied. This is the result of the imaginative manipulation of limited means.

Design number 189 is the richest. It is the only one that includes carving (the furled leaves at each end), and its assembly is unlike that of the others. For these reasons it should be set aside. The remaining seven can be analyzed together. Each bench includes two end pieces. These consist of flat

[43] Gay, "The Furniture Trade," in Chauncey M. Depew, ed., *One Hundred Years of American Commerce,* 2 vols. (New York: D. O. Haynes & Co., 1895), 2:629.

[44] Meeks, *The Railroad Station* (New Haven: Yale University Press, 1956), pp. 1–25.

[45] If not labeled or stamped.

[46] Giovanni Morelli, a nineteenth-century Italian art historian, maintained that one artist's work could be distinguished from another's through the careful examination of the parts of lesser significance in a picture. Because they were less important, an artist would execute them automatically, and certain small but distinctive personal treatments would be repeated. Thus, in furniture, subtleties of machine application might be more fruitful for attribution than more apparent elements of design.

[47] Photographs in the Phoenix catalog are several times inscribed in ink: "Made by Phoenix 1878." The earliest Nelson, Matter and Company album contains a photograph of the firm's award certificate from the Centennial.

Fig. 4. Phoenix Furniture Company, footrests. From an album of Phoenix Furniture Company photographs, 1878. (Grand Rapids Public Library.)

boards—perhaps, though not necessarily, of one piece—which were given elaborate contours by the band saw or jigsaw. A jigsaw was also used to cut out or perforate each of the end pieces, resulting in a more complex system of solids and voids, lights and darks. This piercing created an impression of complex construction. After the contour cutting, a second step was the use of a molder to give more interest to some of the contours and even some of the pierced forms. The third and, for some of the end pieces, the last step was the addition of ornament of two basic varieties: roundels and panels.

The roundel is perhaps the single most common decorative feature of the Renaissance style. Indeed, it is nearly ubiquitous in furniture of this decade. Each of these eight benches has at least two of these motifs (one on each end); number 183 has fifteen, and number 186 has seventeen. Related to roundels are other turned ornaments, usually used as urns, finials, or drops, which appear, in particular, on designs 188 and 189.

Paneling is the other major type of decoration. Again, a simple process was followed to produce a rich-looking effect. Thin panels of wood, usually between ¼ and ½ inch in thickness, were cut into a variety of shapes. The edges were generally given a concave bevel. Burl veneer, cut by veneer saws sometimes to a thinness of $\frac{1}{40}$ of an inch or less, was then glued to the panel. The pieces shown here display several different shapes of panels. Larger

and more elaborate panels will be seen in subsequent designs. A possible fourth step in decorating these end pieces was the incising of delicate linear patterns. This too was accomplished by machine and was an inexpensive way to impart further visual interest to the product.

In addition to the matching end pieces, these benches consist of a few other basic parts. Each has a simple plank seat with a molded wooden frame around it to be padded and upholstered. Some of the stools include compartments for which the seat serves as a lid. The only other element used in the construction of these pieces is a stretcher that runs across the bottom of the stool to strengthen it. These stretchers are either turned pieces decorated with roundels or finials, or shaped boards parallel to the floor or perpendicular to it, again ornamented with roundels.

Through simple methods of construction, an impressive array of designs was created. No two stools look quite alike; each is a fresh solution to the problem within a definite stylistic framework. These construction and design characteristics were neither invented nor monopolized by Grand Rapids firms but were successfully exploited by them. Their application to chamber and parlor furniture will be demonstrated below.

The three basic pieces of a normal chamber suite can be seen in Phoenix's suite number 302 (Fig. 5), a modestly priced example near the bottom of the

Fɪɢ. 5. Phoenix Furniture Company, chamber suite. From an album of Phoenix Furniture Company photographs, 1878. (Grand Rapids Public Library.)

firm's line. Of these pieces, the largest and most important in most suites was the bed. Most beds, like this one, consisted of four parts: headboard, footboard, and two side rails. Typically the headboard was not only more ornate than the footboard but higher as well. Sometimes it was only half again as high, as in this suite; more often it was two or even three times the height of the footboard. The two side rails were usually joined to the endboards with cast iron fixtures of a sort that facilitated the rapid assembling and dismantling of the bed. These cast iron devices replaced the older bolts, which were less satisfactory. Slats were placed across the bed from one side rail to the other, resting in indentations provided for them. On these slats rested the spring mattresses that were popular in the 1870s. The bed belonging to this suite is of a relatively simple design that, typical of less expensive furniture, is somewhat *retardataire*. The curvi-

linear quality recalls the styles of 1860 more than those of the late 1870s, even though the bed has been somewhat updated by the intricate Renaissance motifs glued onto both the footboard and the headboard.[48]

The second essential piece for most chamber suites was a case piece with a mirror. This suite is provided with a walnut chest having three drawers. As is often the case with modestly priced suites, the chest is not actually en suite with the bed; it is similar in style rather than identical in detail. Although the most common bureaus of the 1870s had four drawers rather than three and were furnished

---

[48] Some elaborate beds retained the use of bolts, perhaps because their great weight required considerable support. Horace Greeley et al., *The Great Industries of the United States* (Hartford: J. B. Burr & Hyde, 1872), pp. 503–6.

FIG. 6. Phoenix Furniture Company, chamber suite. From an album of Phoenix Furniture Company photographs, 1878. (Grand Rapids Public Library.)

with pulls instead of relying on a key to bring out the drawers, in other ways this bureau is typical of its price class. The veneered panels on the drawers were common, as were the two small lift-top boxes on the top of the bureau, which could be used for cufflinks, collars, combs, or jewelry. This mirror, itself of a somewhat eccentric shape, is held in place by two standards decorated with panels and roundels and resembling inverted monopodia in shape. The crest on the mirror approximates the crest on the headboard with its vague segmental pediment interrupted by a leafy cartouche.

The third piece accompanying most suites was the washstand. Its form varied little. Generally two doors were placed below a drawer. At the back of the top was a splashboard. Here the top and splashboard are wood; more often they were marble.

A more expensive suite appears in figure 6. The overall impression is of a great increase in richness and monumentality. One of the most striking changes is in the size of the bed. Beds in inexpensive suites frequently were relatively small. In this more costly suite the bed has grown to the height of the tall case piece; the headboard here is about twice the height of the footboard. A comparison of this bed and the last one reveals numerous differences ranging from the subtle to the obvious. A subtle distinction appears in the side rails, which in this example are slightly more elaborate, with a more intricate contour and a strip of molding for greater emphasis. While the footboard is fundamentally the same in construction, the end result is obviously much different. Since the differences point up important distinctions between price classes of beds, they are worth noting. In the less expensive bed, the posts in their wider dimension are parallel to the side rails and have an irregular shape. In the more expensive bed, the posts are

Fig. 7. Phoenix Furniture Company, chamber suite. From an album of Phoenix Furniture Company photographs, 1878. (Grand Rapids Public Library.)

square in section with molding at the top and base. The mortising of the individual boards into the posts is the same, but the more expensive bed achieves its effects through heavier molding and more paneling. In short, more expensive chamber pieces were, as a rule, not only heavier visually, because more ornament was applied, but heavier physically as well, because more wood was used in their construction. What has been said of the footboard applies in a general way to the headboard as well.

The differences between washstands are not great; they rarely were. The shapes of the panels on the doors have changed somewhat and ebonized teardrop pulls have been added to the drawer. Otherwise the most significant addition is the marble top and splashboard with its bracket shelves. The decade of the 1870s had great enthusiasm for marble tops, and, if at all possible, they were provided on case pieces and tables. Marble was used in

a wide variety of colors: white, black, yellow, orange, jasper, and gray. Of these, the most common was white.

The case piece in figure 6 is much different from the bureau in figure 5; in fact, it is an altogether different piece of furniture. Called a dressing case, it is characterized by three horizontal surfaces rather than one and by a large glass to reflect the image of the person standing, or dressing, before it. To an extent, one might suggest that this piece represents a fusion of the functions of the traditional chest of drawers and the old cheval glass, no longer in fashion. The relationship of this piece to a bureau like the one in the previous suite is apparent. The drawers on each side of the mirror are akin to the small boxes at the base of the mirror standards in the less expensive piece, but they are significantly larger. As the mirror expanded, it depressed the carcass of the piece, reducing the number of full-width drawers from three or four to

Our magnificent new Warerooms, Canal St., Grand Rapids, contain a complete and varied assortment of every description of FURNITURE.

Hotel Furnishing a specialty.

[*See Page 17.*]

Fig. 8. Berkey and Gay Furniture Company, chamber suite. From J. D. Dillenback, *Grand Rapids in 1874* (Grand Rapids: by the author, 1875), pp. 56–57.

two. The large glass was not only functional; it was enjoyed for its visual qualities as well. Americans of the 1870s employed large mirrors in pier glasses, sideboards, hall stands, and over mantels, as well as in dressing cases. Not only did the glass reflect and thus multiply the light, but it also reflected picturesquely fractured images of interiors. About halfway up each side of the mirror are small brackets intended to support candles or, more likely, small oil lamps. These provided light for dressing and applying makeup and served as a nineteenth-century prototype of the modern dressing mirror surrounded by incandescent bulbs.

A third chamber suite, Phoenix's number 322 (Fig. 7), is even more elaborate than the second. The bed has grown to magnificent proportions and represents a masterpiece of the furniture maker's art. The headboard is now well over twice the height of the footboard, its great expanse imaginatively ornamented by the simple yet effective means available to the factory designer. The headboard is essentially a manipulation of five superimposed planes. The first plane of the headboard consists of the extensive areas of burl veneer glued to thin boards and screwed onto the back of the piece mor-

tised into the posts at either side of the headboard; this latter piece constitutes the second plane. The third plane comprises the panels and roundels applied to the second. Toward the top of the piece, a fourth plane projects beyond those below; on the projection are superimposed still more applied decorations, making the fifth plane. The advancing movement thus established by these planes contributes greatly to the visual success of the bed.

There is only a modest amount of carved decoration on the bed: a few foliate forms and a cartouche. Otherwise panels and roundels dominate the design. There are two elongated panels flanking a roundel on the side rails, another roundel centered on the footboard and one on top of each of the posts, as well as five roundels (four rosettes and one disk) on the headboard. The panels are even more numerous and easily seen.

Other suites show how consistent were the essentials of this style yet how varied the product. In 1874 Berkey and Gay advertised a slightly more elaborate suite than those above (Fig. 8). While more carving is in evidence, as well as a somewhat more lavish use of panels and roundels, this design adheres to the basic principles followed in the de-

FIG. 9. Berkey and Gay Furniture Company, chamber suite, ca. 1876. From a trade card. (New-York Historical Society, Landauer Collection.)

sign of most Grand Rapids Renaissance furniture. For example, the upper section of a bed was almost invariably designed to draw the eye. It is the over-scaled, jagged quality of these upper sections that makes this furniture so bold and intimidating. Because the design expands actively near the top instead of subsiding quietly, our eyes are drawn to the upper area of the bed. One of the most common devices for achieving this excitement is shown in this design, where a rudimentary pediment or arch is exploded by an ascending cartouche. A different method for drawing attention to the summit of a piece can be seen in another design by Berkey and Gay, perhaps dating from around 1876 (Fig. 9). A heavy molding suggesting an arch moves across the top of the pieces. Because it moves beyond the uprights a matter of a few inches on each side, it suggests the continuation of that line and constitutes an expansive rather than a confined design. In a similar way, the extension of the uprights through this quasi-arch device projects the movement up-

ward, creating that sought-after top-heavy feeling. In addition, the central decorative element of disk, scrolls, and anthemion forms a pointed composition moving upward. All such elements helped to create the awkwardly assertive quality that the public of the 1870s found so pleasant.

At the top of each company's line were productions that came close to the level of exhibition pieces. Nelson, Matter and Company's suite number 112 (Figs. 10, 11) was of a particularly grand design. In rich and expensive pieces like these the real strength of the Renaissance style can be seen. Bold, vigorous, aggressive, such creations represent the antithesis of the calm, quiet, and stability of the Federal style, popular some seventy years before. While it may be difficult for many twentieth-century eyes to find enjoyment in such productions, it is important to remember that few people of the 1870s could find much delight in what they felt to be the boringly tame creations of the Federal period.

FIG. 10. Nelson, Matter and Company, bedstead from a chamber suite. From an album of Nelson, Matter and Company photographs, ca. 1876. (Grand Rapids Public Library.)

FIG. 11. Nelson, Matter and Company, case pieces from a chamber suite. From an album of Nelson, Matter and Company photographs, ca. 1876. (Grand Rapids Public Library.)

Even while departing from the general feeling of the Federal period, this furniture shares with it certain principles of furniture design. Each one of these pieces is designed not only with an eye on the market but with more than one backward glance at traditional ideas about composition. In this giant bedstead, for example, bilateral symmetry is highly important. A central axis, bisecting the footboard and the headboard, is repeatedly emphasized. Starting at the bottom of the footboard, we find it first in the center of the small apron, which stands out so well against the shadow under the bed. Not only does the shape of the apron indicate the central axis, but the delicate incised ornament found there includes a foliate device that might as well be an arrow indicating the way our eyes are to move. The central axis is next emphasized by the leaf that curls from a sliced cartouche over a molding onto a panel of matched burl veneer, then by the modest but apparent jogs at the base of the

shaped panel, and finally by the stalactite of the houndstooth design.

In the headboard the axial organization is continued by the small incised plaque; the roundel above it; the matched veneer panel; another furled leaf, twin to that on the footboard; more shaped panels, relatives of if not twins to those on the footboard; the void between the fifth and sixth of the houndstooth elements; the semiroundel; and, at last, the Greek palmette at the apex. Perhaps this strong axial symmetry is necessary to withstand the assaults of less traditional composition devices: the spiky finials along the top of the design that tend to pull and extend it upward; the acroterionlike devices that project the design outward and upward at a 45-degree angle; and lastly the nearly overpowering scale of the top element, an inventive version of a temple front.

The two case pieces with this bed are somewhat different from any discussed so far. The smaller one is a somno, or nightstand, a piece usually found only with more expensive suites. The larger piece is another version of the dressing case, with a stunted base and a greatly expanded glass, but without the small drawers at the side.

Expensive suites were occasionally furnished with two varieties of dressing cases (Figs. 11, 12). Here the storage space in each of the dressing cases is rather limited, especially in the one with the depressed center (Fig. 12), a magnificent design wholly of the 1870s. The bold contrast between the three small drawers on the sides and the great dip in the center was exactly what the period enjoyed. Clearly, both of these pieces of furniture were more important as decorative mirror cases than they were for storage; clothing was placed either in wardrobes or in closets, while the walnut luxury of the chamber furniture was reflected in great expanses of glass.

One of Phoenix Furniture Company's original purposes was to manufacture upholstered parlor furniture, for William Berkey realized that there was a great potential market for such a product.[49] Thus it is not surprising that the most numerous and impressive pieces of furniture in Phoenix's 1878 catalog belong to this category. Like chamber furniture, parlor suites demonstrate a considerable range in price and elaboration. And again like chamber suites, cheaper designs often reflect earlier styles updated through the addition of fashionable details.

A useful suite with which to begin is Phoenix's parlor set number 112 (Figs. 13, 14). There is nothing particularly remarkable about this ordinary, serviceable set of furniture from near the bottom of the firm's price range. Common in the 1870s, such furniture can still be had today at a modest cost. To appreciate the essentially conservative nature of these pieces, it is necessary to examine them in some detail.

Although produced in the 1870s, the sofa represents a design concept at least a generation old. It is fundamentally a version of the small curved-back rococo-style sofa common in the 1860s. This, in turn, had been based on a transitional design—late classical to rococo—of the 1850s. The sofa retains its rudimentary cabriole legs, cut by band saws and curved only in two dimensions, not in the full round. The arm supports and front legs are formed from one continuous piece. A slight jog is visible on

the outside of each arm support, near the top. It is not difficult to see this as the remnant of the swan's beak that might have been there when the design originated under the declining influence of the Empire style. No other trace of a swan appears now; thin veneer panels have been applied to the armrests, and a roundel appears where a swan's eye once might have been. These same familiar devices have been applied to the back to bring it up to date. Six panels of veneer of three different patterns alternate with five roundels of three different sizes and two different patterns.

The chairs that accompany this suite similarly represent the updating of earlier forms. In this case the prototype is the upholstered, tall-back chair known as the Voltaire, which was popular in France during the Restoration and appeared in America about 1830. While the Voltaire possessed a modest amount of grace, it was of a rather rigid design. If anything, these descendants are even more rigid. Their backs are completely flat, with no concavity. Although the seats are somewhat wider at the front, and the seat rails curve gently, these chairs are almost totally rectilinear in construction. Undulating shapes and applied curvilinear ornament disguise this to a degree but do not hide it. The joinery in these chairs, unlike that in more expensive pieces, is direct, with little subtlety or subterfuge about it. The gentleman's chair, for example (Fig. 14, center), is made from eleven pieces of wood cut on band saws and molded. Most pieces are of solid walnut, but those that form the seat are more likely walnut veneer over a less valuable wood. Every piece used in the construction of this chair is flat on two sides, revealing its derivation from a board. It is not difficult to reconstruct the processes leading to the creation of this piece: felling the timber, cutting it into boards and seasoning them, applying templates, and cutting the shapes. The lady's chair and the side chair contain even fewer pieces. Thus frames for the entire suite —sofa, lady's chair, gentleman's chair, and four side chairs—could be manufactured quickly, because they made efficient use of a few simple machine operations, and inexpensively, because relatively little time went into their fabrication and material was not costly.

The unpretentiousness of this suite can be fully appreciated when it is set beside an obviously much more expensive but somewhat similar one (Figs. 15, 16). Perhaps what is remarkable about these two suites is not their similarities but their differences.

---

[49] Ransom, *City Built on Wood*, p. 14.

FIG. 12. Nelson, Matter and Company, case pieces from a chamber suite. From an album of Nelson, Matter and Company photographs, ca. 1876. (Grand Rapids Public Library.)

Although the proportions and construction are much the same, the second seems far removed from the first. This great difference results from the cumulative effect of a series of relatively minor changes. The rudimentary cabriole legs are gone, replaced by tapering, turned legs, more appropriate to this style. Flutings and incised designs are gilt, adding a degree of elegance and luxury to the sofa. There is a tendency to fracture the design by breaking its continuity. In the conceptually older piece, the sides flow gently into the back; in the newer one, the arms curve less, and there is a notable hiatus between the sides and the back. This interruption serves to set off the individual units much more strongly and to provide that sense of discrete elements typical of much furniture of the 1870s.

There are other ambitious design elements that distinguish this sofa from the cheaper version. The apron is delicately shaped and pierced. The regular, strict rectangle of the back is given life by the way the vertical and horizontal rails cross at the upper corners, and by the fine crest with its medallion portrait, derived from French chairs of the early seventeenth century. The finest features of all are the carved caryatids supporting the arms. This

device is widely found throughout the period on high-quality furniture made in most production centers.

One last distinction may strike the eye of the viewer at once: the upholstery on the second set is far more elaborate. It is curious that in the nineteenth century, when the machine threatened to reduce all furniture to the same price range and thus rob it of its status-conferring function, hand labor intervened to protect the concept of hierarchy. Because upholstery was done by hand, it was expensive. The more elaborate the upholstery, the more costly. The modest sofa used somewhat less than 50 buttons in its upholstery; the rich one used nearly 125. This, of course, not only demanded more time and fabric but resulted in a more luxurious creation. There is something crude and dowdy about the first suite when it is placed beside the second.

With the first suite it seems that the flat back is used simply as a matter of economy. The presence of flat backs in the second suite makes it clear that their use constitutes essentially a stylistic device. The style, derived from both seventeenth-century and Louis XVI sources, was advantageous to those who wished to produce cheap furniture, since straight units were easier and cheaper to make than curved ones. In moderate to expensive pieces, backs were more often curved than straight, but straight backs could be found in both inexpensive and costly designs.

Mention has been made of the late-nineteenth-century interest in creating a composition in which the various parts, while subordinate to the whole, demonstrate a strong degree of individuality. This tendency, visible in most furniture of the period, including the chamber furniture discussed above, is particularly apparent in sofas. The unified (or near unified) backs of earlier designs were never as popular as their more elaborate alternatives. In fact, the most characteristic form of sofa in the 1870s was a triple-back variety. The sofa in figure 17 gives a fine example of the visual tension so much enjoyed at the time. The three panels of the back, with carefully corresponding crests and details, seem to be caught in an explosive movement away from the core of the design, a movement similar to that observed in chamber furniture. To produce this effect, the joinery has become quite intricate. Between the upholstered panels and the outside supports of the frame are pierced areas that exaggerate the delicacy of the design. The unusual

FIG. 13. Phoenix Furniture Company, sofa from a parlor suite. From an album of Phoenix Furniture Company photographs, 1878. (Grand Rapids Public Library.)

FIG. 14. Phoenix Furniture Company, chairs from a parlor suite. From an album of Phoenix Furniture Company photographs, 1878. (Grand Rapids Public Library.)

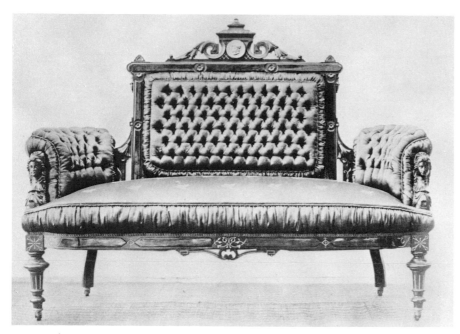

FIG. 15. Phoenix Furniture Company, sofa from a parlor suite. From an album of Phoenix Furniture Company photographs, 1878. (Grand Rapids Public Library.)

FIG. 16. Phoenix Furniture Company, chairs from a parlor suite. From an album of Phoenix Furniture Company photographs, 1878. (Grand Rapids Public Library.)

Fig. 17. Phoenix Furniture Company, sofa from a parlor suite. From an album of Phoenix Furniture Company photographs, 1878. (Grand Rapids Public Library.)

Fig. 18. Phoenix Furniture Company, chairs from a parlor suite. From an album of Phoenix Furniture Company photographs, 1878. (Grand Rapids Public Library.)

Grand Rapids Furniture                                                                                           47

ovals capped with stylized palms between the upholstered panels contribute significantly to the agi-
tation of the design.

The means used to unify this expansive composition should be noted: first, the application of the
principle of symmetry; second, the repetition of
shapes and designs; and, third, the utilization of
what amounts to a belt course to tie the entire piece
together. The arms, beginning at the front, establish a horizontal band that passes around the entire
piece, holding the panels together. While the horizontal line is interrupted by the panels of the upholstery, it reappears in the two ovals to maintain
a state of peaceful coexistence among the various
design elements.

As is usually the case, the chairs accompanying
this suite (Fig. 18) are less bold in design than the
sofa, which was usually the *pièce de résistance* of
a parlor suite. They do, however, reveal the characteristic differentiation of types. Parlor suites generally included three types of chairs: a side chair,
a lady's chair, and a gentleman's chair. The side

chair (left) was the smallest. It had only rudimentary "bracket arms," as they were called, but no
arms in the usual sense. Its back was the lowest of
the three types, and its seat the narrowest and the
greatest distance from the floor. The lady's chair
(right) usually had low arms to accommodate the
elaborate dresses of the day. The gentleman's chair
(center) was the largest and boasted full-sized arms
and the widest seat.

Another version of the triple-back sofa appears
in figure 19. The last sofa (Fig. 17) was partially
unified by the repetition of motifs with a variation
in scale. We might describe the arrangement of the
upholstered panels as a-A-a. Here, on the other
hand, is a system of B-A-B, for the two flanking
panels are entirely different in shape from the one
in the center. The central panel is oval; the flankers are wedge-shaped. The central panel culminates
in a grand bit of carving; the flankers have curled,
upholstered tops. This is the sort of contrast the
Victorian eye found most enjoyable.

The joinery on this suite is particularly involved,

FIG. 19. Phoenix Furniture Company, parlor suite. From an album of Phoenix Furniture Company photographs, 1878. (Grand Rapids Public Library.)

FIG. 20. Phoenix Furniture Company, sofa from a parlor suite. From an album of Phoenix Furniture Company photographs, 1878. (Grand Rapids Public Library.)

especially in the way the central panel is joined to its flankers. Complicated joinery can also be seen in the deliberately bizarre treatment of the arm supports of both the sofa and the gentleman's chair. A curved piece rises from the seat. In a tortured S-curve it gradually wends its way up to the arm, but halfway along its length it supports a plump turning, which also rises to the arm. Something here is superfluous, but it is not clear exactly what.

Less common than either the single- or the triple-back sofa was the double-back sofa. One particularly pompous example appeared in Phoenix's catalog (Fig. 20). It deserves attention, for it is a most unusual design, with many uncommon details. First, it has three legs across the front rather than two. Second, because there are only two upholstered panels in the back, an elaborate piece of woodwork occupies the center of the design. Third, the upholstered panels of the back are circular, which is exceptional, and they appear more or less

wedged between spreading uprights. Fourth, the upholstery lavished on this bizarre frame is particularly exuberant. While the sofa is shown only in its muslin undercovering, the chairs accompanying it (Fig. 21) give a better indication of the appearance of the finished product, with thick padded upholstery oozing luxuriously over the frames.

The Grand Rapids products surveyed here have demonstrated repeatedly the characteristics of furniture design in the 1870s. The best design was bold and vigorous; subtleties were avoided because they were uninteresting. Novelty within a broad stylistic framework was highly valued; and the stronger and more emphatic a design, the more popular it was likely to be. Wood was varnished to produce a smooth, gleaming surface. Upholstery fabric was rich and colorful, and upholstery techniques often differed on a single piece for the sake of variety. Yet while Grand Rapids excelled in all these stylistic touches, none of them was unique to the city. The concepts, the designs, and the tech-

Fig. 21. Phoenix Furniture Company, chairs from a parlor suite. From an album of Phoenix Furniture Company photographs, 1878. (Grand Rapids Public Library.)

niques were shared by furniture concerns across the country. Grand Rapids prospered, not because of the invention of these designs, but because of their application.[50]

Throughout this discussion, occasional assertions have been made about the reasons for the success of Grand Rapids. At this point it is appropriate to consider this matter at greater length. Tangible assets certainly were only partly responsible, as William Widdicomb found when he addressed this question more than half a century ago. It is true that lumber was abundant in the area, but it was equally abundant throughout much of the North. Waterpower was also available, but this too was found throughout the North. One could just as easily turn this kind of argument around and assert that Grand Rapids had no special natural advan-

tage over the rest of the North. In fact, it might even have been at something of a disadvantage: at the outset it was served by only one railroad, which terminated at Lake Michigan on one side and Detroit on the other. As Widdicomb noted:

When, eventually, we did have a connecting railroad with the Michigan Central and Lake Shore & Michigan Southern our whole product was freighted through towns where many well-established competitors were located. Upon the Michigan Central were Buchanan and New Buffalo, both manufacturing upon a larger scale than ourselves. Upon the Lake Shore were to be found La Porte, Mishawaka and South Bend, each having one or more successful furniture factories. Chicago was the distributing point, and there were, as at present [1909], other and stronger competitors, yet the city of Grand Rapids rapidly passed all of them.[51]

Examination of census reports makes it clear that Boston was also an important center of furniture

[50] A description of the English version of this taste, termed *parvenu*, can be found in Winslow Ames, *Prince Albert and Victorian Taste* (New York: Viking Press, 1968), pp. 184–85.

[51] Widdicomb, "Early History," p. 74.

manufacture. In addition to the usual general line, it was known for the production of fine chamber furniture. It gave birth in 1870 to America's first furniture journal, *The Cabinet Maker,* which was national in scope. When Boston firms failed, Bostonians typically blamed the ruinous competition from Grand Rapids and other midwestern cities, yet these Boston firms "had all the advantage in prior possession of the field, abundant capital, fine factories, and a near location to the market."[52]

For Widdicomb it was nothing so easily quantifiable as lumber, water, railroads, or location that led to success, but rather the personality and talents of the men who founded the Grand Rapids firms. Perhaps this, in the end, is the real answer. Certainly every large success is the result of a number of forces at play; but, without a doubt, the men who started the great enterprises of Grand Rapids deserve more than a little credit for their ultimate success. Although they were not inventors like Henry Ford or Thomas Edison and were never as successful as Andrew Carnegie or John D. Rockefeller, they were nonetheless of a similar breed. They realized early in their careers the importance of sophisticated marketing techniques. When Europeans visited the Centennial in Philadelphia, they were impressed not only by the ability of Americans to produce attractive cheap merchandise by utilizing laborsaving machines and mass production but also by their skill in packaging and marketing products.[53] The industrialists of Grand Rapids early employed marketing techniques that have become commonplace today.[54] Perhaps their greatest

achievement was the institutionalizing of the semiannual furniture market, which for years brought hundreds of buyers to the city. Through sophisticated marketing and what must have been a determined effort to create a reputation for quality, these men made the small city of Grand Rapids known, not only throughout this nation but abroad as well, as "the mother of art and comfort." [55]

To an extent one might be able to argue that the glorious days of Grand Rapids coincided with a particularly expansive period in the history of American furniture. The ancient trade of furniture making grew throughout the nineteenth century; the 1860–70 decade, when national production of furniture increased two and one-half times, was especially impressive. After that date growth began to level off. While per capita expenditure for furniture had been $1.77 in 1870, in 1880 it declined to $1.55, and by 1890 it had further diminished to $1.38.[56]

Grand Rapids remained an important center for the production of furniture in America into the twentieth century, but the depression struck it a blow from which it never fully recovered. The newer factories in North Carolina, with their lower labor costs and low- to medium-priced lines, suffered less and now prosper, while Grand Rapids is but a ghost of its former self.[57]

---

[52] Widdicomb, "Early History," p. 74.

[53] Maass, *Glorious Enterprise,* pp. 104–12.

[54] For an indication of the complexity of marketing in to-

day's furniture industry, see the discussion of the Drexel Furniture Company in Milton P. Brown, Wilbur B. England, and John B. Matthews, Jr., *Problems in Marketing* (New York: McGraw-Hill, 1961), pp. 401–32.

[55] Gay, "Furniture Trade," p. 632.

[56] Gay, "Furniture Trade," pp. 630–31.

[57] David N. Thomas, "A History of Southern Furniture," *Furniture South* 46, no. 10, sec. 2 (Oct. 1967): 59, 72.

# The Dover Manufacturing Company and the Integration of English and American Calico Printing Techniques, 1825–29

*Caroline Sloat*

SEVERAL NEW ENGLAND textile companies competed during the 1820s to establish the first large-scale textile printing operations in the United States. Firms in Dover, New Hampshire, and in Taunton, Chelmsford (later Lowell), and Fall River, Massachusetts, committed large sums of capital and years of effort to mastering the techniques of calico printing. They developed printworks capable of handling the thousands of yards of cloth spun and woven in their cotton mills, as well as cloth from other New England mills. As a result of their efforts, large quantities of domestic printed cottons became available in the following decade in even the smallest American towns.

Boston agents to whom these new printed goods were consigned sold them to merchants from urban and country stores. Newspaper advertisements soon began to include descriptions of "good assortments of calico" and "Turkey Red and other prints." A store in Colrain, Massachusetts, advertised "100 ps New style fancy prints" as part of its 1828 spring stock. The quantities of printed textiles that eventually became available are also indicated in the inventories of country stores. For example, Asa Knight of Dummerston, Vermont, had 1,688 yards of calicoes on his shelves in 1846, part of a textile inventory that included sheetings and shirtings from several New England factories, as well as woolens, silks, and linens. At his death, in 1851, Knight had 2,036 yards of prints alone in stock.[1]

The availability of printed textiles and their in-corporation into daily use for clothing and interior decoration in Victorian-era households illustrate one facet of the taste of the early part of this period. Among his recommendations for the treatment of cottage interiors, Andrew Jackson Downing mentioned window curtains as "bestowing an air of taste and refinement." He suggested that both chintz and printed cotton, in addition to the more costly moreen, might contribute to this effect. Moreen, a woolen fabric commonly used for curtains and bed hangings, was losing favor because of the dust and moths it collected. The authors of *An Encyclopaedia of Domestic Economy* felt that "chintz is generally preferred being more easily washed" and more healthful than moreen for curtains. Chintz was also recommended as a covering for couches in bedchambers and ladies' sitting rooms. Calicoes were extensively used for items of clothing, shawls, and handkerchiefs, while scraps found their way into patchwork quilts and even doll's clothing. "Most girls begin dress making very early," noted the author of a guide for seamstresses, "that is to say, they clothe their dolls . . . when their mother gives them a piece of print."[2]

The manufacture of these prints in such quantities was a triumph over adverse conditions. Until the early 1820s there was no capacity for large-scale textile printing on the American continent. John Hewson and other block printers had worked in the Philadelphia area since the Revolution, and a printworks in Charlestown, Massachusetts, pro-

[1] I would like to acknowledge the assistance of my Old Sturbridge Village colleagues Richard Parks, Richard Candee, and Theodore Z. Penn in the preparation of this article. *Greenfield Gazette* (Greenfield, Mass.), May 19, 1828; Asa Knight, Dummerston, Vt., Invoice of Goods, Apr. 1, 1846, Schedule of Goods in Store, Aug. 19, 1851, Asa Knight Estate, Probate Court, District of Marlboro, Brattleboro, Vt.

[2] Downing, *The Architecture of Country Houses* (New York: D. Appleton & Co., 1853), pp. 373–74; T. Webster and Mrs. Parkes, *An Encyclopaedia of Domestic Economy . . .* , ed. D. M. Reese (New York: Harper & Bros., 1848), p. 291; *The Ladies Self-Instructor in Millinery, Mantua-making and All Branches of Plain Sewing* (Philadelphia: G. B. Zieber & Co., 1845), pp. 23–43.

duced small quantities of block prints during the mid-1820s. But neither these early firms nor the cylinder printeries established around Philadelphia in the early nineteenth century challenged the flow of imported prints or the supremacy of British printing technology.

Several New England firms, larger and better financed, soon dwarfed these early efforts. Among the three earliest was the Taunton Manufacturing Company, incorporated in 1823, which combined under one management several preexisting cotton factories, much technological expertise, and the financial backing of wealthy Boston merchants. Another, the Merrimack Manufacturing Company of Chelmsford, had been formed to build the locks and canals at Lowell, as well as textile mills to produce cloth for printing. The Dover Manufacturing Company, like the Taunton firm, represented the expansion of an earlier corporation into a well-financed business. These early firms were on familiar ground in constructing large factory buildings to house their new printing operations, or at least they had all of the resources needed to solve construction problems. But all of the companies soon discovered enormous gaps in their knowledge of the mechanical and chemical technology of calico printing. Determined to succeed, their directors quickly set in motion a scramble for technological competence that soon led them to rely on British expertise in machine building, chemistry, and design. In fact, their achievements in printing were due in large part to their success in acquiring British skills and equipment.[3] The style and quality of American printed textiles was inevitably the subject of continuous comparison with contemporary British prints, which were produced under more favorable circumstances. But by 1853 two British experts traveling in the United States, fully aware of the dependence of the American industry on British technology and manpower, granted that "American printed goods evince progress of no ordinary kind." [4]

The entry of the Dover Manufacturing Company into textile printing is well documented in three surviving company letter books that cover the years 1825–29. Workbooks compiled by two men who actually printed calicoes at Dover provide further information about printing skills as practiced and taught by Englishmen brought to America by the company. In addition, Dr. Arthur Livermore Porter, the first chemist at Dover, published an American edition of Samuel Frederick Gray's *Operative Chemist* after he left the printworks, in 1828. Entitled *The Chemistry of the Arts; Being a Practical Display of the Arts and Manufactures Which Depend upon Chemical Principles,* this book provides further insights into early practices at Dover. Porter's most important addition to Gray's work was a new section, "Treatises on Calico Printing and Bleaching," based on his Dover experiences. Shortly after he was hired, the company expanded both its physical and its technological capacities. Porter was largely responsible for assembling a work force that understood the chemistry and applications of both block and cylinder printing. Gaining this expertise was not as simple as merely importing British workers, for immigrants had difficulty adapting their techniques and work styles to the American situation. The well-documented operation at Dover is an important example of the difficulties of technological transfer between the United States and Great Britain.

The first textile company in Dover, the Dover Cotton Factory, had been chartered by the New Hampshire legislature in December 1812. Its wooden building had been constructed at a site on the Cocheco River that came to be called the "upper factory" when, in 1822, the company launched a massive construction program at the lower falls in the river.[5] The downstream site had been acquired to implement ambitious manufacturing plans backed by a $1 million capitalization and a new name, the Dover Manufacturing Company. Two brick mills for the manufacture of cotton cloth were built at the new site in 1822 and 1823. Both were 155 by 43 feet, the first (number two) of four stories and the second (number three) of five stories plus an attic. In construction, as in timing and investment, the development of Dover paralleled the

---

[3] See Theodore Z. Penn, "The Introduction of Calico Cylinder Printing in America: A Case Study in the Transmission of Technology," in *Technological Innovation and the Decorative Arts,* ed. Ian M. G. Quimby and Polly Anne Earl (Charlottesville: University Press of Virginia, 1974), pp. 235–55.

[4] Nathan Rosenberg, ed., *The American System of Manufactures: The Report of the Committee on the Machinery of the United States 1855 and the Special Reports of George Wallis and Joseph Whitworth 1854* (Edinburgh: University Press, 1969), p. 247.

[5] D. Hamilton Hurd, comp., *History of Rockingham and Strafford Counties, New Hampshire, with Biographical Sketches of Many of Its Pioneers & Prominent Men* (Philadelphia: J. W. Lewis & Co., 1882), p. 818.

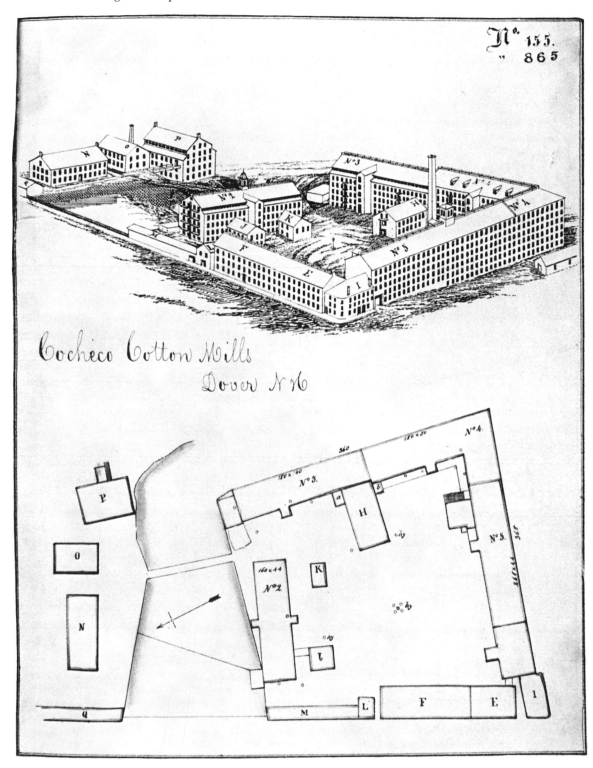

FIG. 1. Isometric view and plan of Cocheco Cotton Mills, Dover, N.H., ca. 1860–75. Insurance Survey, MS. owned by Factory Mutual Engineering Corporation, Norwood, Mass. (Photo, courtesy Randolph Langenbach.) Earlier, in an 1827 listing for insurance purposes, number four was called "manufacturing part of large mill," and number five "printing part of large mill." The buildings labeled *E* and *F* probably are the "building on the street including steam boilers, padding room, cloth room, block cutters room, designing offices &c." John Williams to William Shimmin, 1827, vol. 2, Dover Manufacturing Company MMS, New Hampshire Historical Society, Concord, N.H.

founding of Lowell and the construction of new factories there, although Dover was essentially six to twelve months ahead of Lowell in designing machinery and buildings. The mills in both Dover and Lowell shared several characteristics with the newest mill in Waltham (built in 1820). These included a cupola placed at the center of a clerestory monitor roof, which lighted the attic stories, and an integral tower or "porch" (13 by 12 feet), which placed the stairs and service doors outside the main structure (Fig. 1).[6]

Not until the early months of 1825 did the Dover Manufacturing Company begin to develop plans for calico printing, an enterprise that would survive to the twentieth century as the Cocheco printworks. The first necessity was adequate physical structures to house the new printing enterprise. The next Dover mill (number four), begun at the new site in 1825 and constructed during the next three years, was a combination cotton mill and printing works. Described as "the huge works that we are preparing for a printing establishment," it differed substantially from the earlier mills in scale and design and incorporated several important innovations in its construction. When first built, it was larger than any of the early buildings at Lowell. It formed an L measuring 167 feet along the river and was six stories in height above the basement, with an attic lighted, not by the usual windowed clerestory, but by skylights set into the roof (Fig. 2).[7] This form, the scale of the building, and the skylighting of the attic story had never been attempted before in the American textile industry.

Together with a new bleachery and mill number two, mill number four formed a quadrangle enclosing an area of several acres. A quadrangular layout seems to have been typical in early calico factories in the United States, perhaps because of suggestions from English workmen brought over to work in spinning and weaving factories as well as in printworks. By the time Dover had begun its planning, the Taunton Company was already constructing a printworks on a quadrangular plan. Zachariah Allen, a Providence, Rhode Island, mill owner, described a visit to Taunton in March 1824. "The calico printing establishment," he noted in his diary, "is built of brick in a quadrangular form." But the scale of Taunton's buildings in no way matched that of Dover's. At Taunton, "a part of one side of the building is two stories and the remainder one story high,"[8] while Dover facades ranged from three to six stories (Figs. 3, 4).

As construction proceeded, the company searched for the technical resources necessary to a printing establishment: workmen skilled in all phases of the operation, usable print designs, and machinery. John Williams, the resident agent, and his assistant, Matthew Bridge, were primarily involved with the operation of the already existing cotton and nail factories and with details of the construction of the new factory. The work of bringing the needed resources together was assigned to the chemist, Dr. Porter, who had been "engaged . . . to take charge of the chemical department" sometime before mid-July 1825. Porter held a medical degree from Dartmouth College and had studied in Edinburgh before going to the University of Vermont as professor of chemistry and pharmacy. How he was introduced to the Dover management is not known, but since all of his experience had been in teaching medical students, he needed very much to familiarize himself with calico printing techniques. Porter asked William Shimmin, the company treasurer, who conducted the correspondence from Boston, to send several chemical texts to Dover, includ-

[6] Besides the obvious architectural similarities between the Dover and Lowell mills and the Boston Manufacturing Company mills in Waltham (erected 1814, 1818, and 1820), the Boston Manufacturing Company granted the Dover firm use of its machinery patent rights. They also gave John Williams, agent of the Dover Cotton Factory, access to examine the buildings and "such information as they may possess and as may be needed respecting said building." A similar contract granting information was made between the Boston Manufacturing Company and the Merrimack Manufacturing Company in 1822. Indenture, Oct. 12, 1821, vol. 187, Boston Manufacturing Company MSS, Baker Library, Harvard Business School (hereafter BMC); Directors Records, Feb. 8, 1822, BMC.

[7] John Williams to George Bond, July 16, 1825, vol. 1, Dover Manufacturing Company MSS, New Hampshire Historical Society, Concord, N.H. (hereafter DMC).

[8] Diary of Zachariah Allen, 1821–24, entry for Mar. 20, 1824, Zachariah Allen Papers, Rhode Island Historical Society, Providence, R.I.; *Manufacturer's and Farmer's Journal* (Providence, R.I.), Dec. 4, 1823. For examples of English quadrangular plans, see Jennifer Tann, *The Development of the Factory* (London: Cornmarket Press, 1970), pp. 17, 20, 22, 24. Besides Taunton and Dover, the Hudson Calico Printing Company, Stockport, New York, built between 1826 and 1830, formed a quadrangle with its buildings ranging in height from one to four stories and with its secondary buildings within the square. See George S. White, *Memoir of Samuel Slater, the Father of American Manufactures; Connected with a History of the Rise and Progress of the Cotton Manufacture in England and America* (Philadelphia, 1836), p. 402.

FIG 2. Lower bridge and factories, Dover, N.H., ca. 1830. Thomas Edward, Senefelder Lithograph Co., Boston. (Photo, Merrimack Valley Textile Museum.)

FIG. 3. Central Street bridge, Dover, N.H., 1870s. Stereopticon view into mill yard from point between number two and *M* of figure 1. (Courtesy, Society for the Preservation of New England Antiquities, Boston: Photo, O. H. Copeland.)

ing "Bancroft on Colours and Berthollet on Bleaching" and "a small treatise on dyeing." [9]

Porter also began to search for skilled workmen and to study the arrangements of printworks in the United States. He first visited Philadelphia and Baltimore to look at bleaching and printing operations there, but he was unsuccessful in his efforts to hire anyone. Shortly after his return, a Taunton machine builder, John Thorp, Jr., advertised in the *Manufacturer's and Farmer's Journal,* published in Providence, that he had cylinder printing equipment and carved printing blocks for sale, and also that he wished to establish a small machine-building operation. Thorp had recently left his position as superintendent of the Taunton Manufacturing Company's printworks.[10] He and his father visited Dover not long afterward and were shown the block printing facility that had been set up at the upper factory. Highly critical of what they saw, the two men claimed that with their expertise they could conduct a more successful enterprise. Their claims and lofty manner did not impress either Williams or Porter, nor did their subsequent failure to send workmen as promised. The blocks they did send arrived after a long delay but were returned as unsatisfactory.

Although the Dover managers had reacted to the Thorps mainly on personal grounds, a new alternative soon developed that freed them of any need for further dealings with the Taunton machine builders.[11] Aaron Peaslee, a machine builder whose contract with the Merrimack Manufacturing Company had expired, sought work at Dover, and John Williams felt that there was a definite advantage to be gained by hiring someone directly from the com-

FIG. 4. Small section of the printery (facing street), the only surviving portion of the original architectural fabric, Dover, N.H., 1973. (Photo, Peter Randall.)

petition. "Although the Merrimack Co. may have no important secrets to disclose and maybe and doubtless are themselves in the infancy of printing, yet there cannot be a doubt but the possession of their skill and information would be cheaply purchased at [$]20,000. By securing Peaslee we secure both their skill and information in a very considerable degree." Peaslee worked through the winter and built a printing machine that was ready for testing in April 1826, once some necessary parts purchased in England had arrived and were fitted. Ten pieces of cloth were run at the first trial, but it was decided that the machine was not useful for general production.[12] After the failure of this machine, the Dover Manufacturing Company moved toward dependence on British machine technology.

As has been noted, the company had set up block printing facilities at the upper factory before the large new building was ready. Since block printing depended on skilled handwork rather than complicated machinery, the company hoped to start printing as soon as workmen could be found. This would enable the firm to have prints ready for

[9] Williams to Bond, July 16, 1825, vol. 1, DMC; Order 19, in Williams to Shimmin, letter 85, Aug. 10, 1825, vol. 1, DMC. See Edward Bancroft, *Experimental Researches Concerning the Philosophy of Permanent Colours; and the Best Means of Producing Them, by Dying, Callico Printing &c* (London: T. Cadell & W. Davies, 1794); an American edition was published in Philadelphia in 1814. See also Claude-Louis Berthollet, *Elements of the Art of Dying, Containing the Theory of Dying in General, as Far as It Respects the Property of Colouring Substances,* 2 vols. (Paris: Firmin Didot, 1791); an English translation was available from the last decade of the eighteenth century.

[10] *Manufacturer's and Farmer's Journal,* Aug. 1, 1825; Penn, "Calico Cylinder Printing," p. 245.

[11] Williams to Shimmin, letter 99, Sept. 9, 1825, Porter to Shimmin, Sept. 9, 1825, Williams to William Payne, letter 102, Sept. 16, 1825, Williams to Samuel Torrey, letter 103, Sept., 1825, Williams to Porter, Oct. 15, 1825, vol. 1, DMC.

[12] Williams to Torrey, letter 123, Nov. 5, 1825, vol. 1, DMC; Samuel Dunster to ?, May 13, 1880, Historical Memoranda 3, no. 394, Dover Public Library, Dover, N.H.

market long before the establishment of the cylinder printing operation. Porter had been searching for printers, but by November 1825, when four printing tables were ready, only one printer had been hired. As the Thorps had not sent the workmen they had promised, thought was given to advertising for three or four calico printers and a block cutter. Apparently this was done, and shortly afterward the company hired Alexander Rogers "on the terms which the Taunton Company pay their printers." [13] Peaslee also suggested possible workers, and duplicate letters were addressed to Fall River and Taunton in efforts to locate them.

At the end of the year Williams reported, "We are now ready for printing, but as luck would have it our blockprinter got severe injury. [He] will be able to work in a few days." Dover's first block prints were made early in January 1826, so Williams was able to state in his annual report to the directors that the printers "at Upper Establishment are making a beginning and a specimen is herewith furnished." [14] Partly as a consequence of the hiring of Thomas Greenhalgh, an English printer from Lancashire, as supervisor of the printing establishment, more printers were employed during the next few months. Greenhalgh's connections with other English printers in America were an immediate advantage to Dover. "[He] is well acquainted with Mr Yates of Taunton works and many valuable work people who left England about the same time as he did, who would probably go direct to Fall River, they having heard more of that place than any other establishment." [15] Greenhalgh visited Taunton and Fall River, both to search for workmen and to find out how the rival companies operated, "the number of vats, wash wheels, . . . and what quantity of work they are doing at each place and what they expect to do within one year." [16] Greenhalgh's success in recruiting resulted

in an indignant letter from the agent at Fall River protesting the hiring of an English block cutter.[17] For the next year each successive monthly payroll at Dover showed the addition of about three or four men to the work force in the printing establishment.

Greenhalgh's efforts produced modest success for the fledgling printworks. His first prints were sent to Boston with an apologetic accompanying letter from Bridge, the assistant agent. "I trust you will discover considerable improvement in the next prints you receive," Bridge wrote at midyear. By year's end, Williams was able to report that "Greenhalgh is doing us good work . . . and Capt. Paul [Moses Paul, superintendent of the upper factory] appears as well pleased with him as usual, perhaps a little better." [18] Still, the slow pace and other production difficulties forced the Dover Manufacturing Company to conclude, as Merrimack and Taunton had already done, that a major effort was necessary to obtain dependable British machinery as well as skilled British workers. Each company sent representatives to England to learn firsthand how factories were managed, and each firm hired professionals from England to put their printeries in operation.

Taunton, Dover, and Merrimack each sent a ranking employee to England to observe printworks and to recruit workers from the textile centers in Lancashire. The Taunton Company sent Charles Richmond in 1825. Richmond, their agent, had long been involved with the mechanics of textile manufacturing and had visited England before, making some important acquaintances. Dover considered sending its machine builder but then decided on Porter. The chemist left in 1826 and expected to spend six months in Europe, but the visit stretched to more than a year, during which he met manufacturers and druggists from whom he gained ideas, techniques, and supplies. The Merri-

[13] Williams to Rogers, Jan. 2, 1826, vol. 1, DMC.

[14] Williams to Shimmin, Dec. 14, 1825, vol. 1, DMC; Annual Report to the Directors, letter 143, Jan. 21, 1826, vol. 1, DMC.

[15] Williams to Shimmin, letter 34, May 5, 1826, vol. 1, DMC. Andrew Robeson, a native of Germantown, Pennsylvania, came to Fall River from New Bedford, Massachusetts, in 1824. He leased a building to begin a printworks and soon began construction of the Fall River Print Works, which had a long history.

[16] Williams to Greenhalgh, May 13, 1826, vol. 2, DMC. Although building four was structurally complete, no final

plans had been drawn up for the allocation of space or the arrangements for bleaching and printing machinery and equipment.

[17] Williams to Shimmin, letter 42, May, 1826, vol. 2, DMC. The block cutter was Thomas Roberts, who had worked with Greenhalgh in England and who first appeared on the monthly payrolls on May 27, 1826. Pay Roll, Apr. 26, 1826–Jan. 1829, Dover Manufacturing Company MSS, Baker Library, Harvard Business School (hereafter DMC Harvard).

[18] Bridge to Shimmin, letter 47, June 14, 1826, Williams to Shimmin, letter 120, Dec. 4, 1826, vol. 2, DMC.

mack Company sent its agent, the engineer Kirk Boott, in 1826. Merrimack had delayed hiring the highly skilled and well-recommended Manchester calico printer, John D. Prince, until it was actually ready to begin production. The Appletons, the managing family at Merrimack, had known about Prince at least as early as 1822. "We have taken into consideration the question of engaging Mr Prince and are inclined to the opinion that it may or will be deferred for the present for the following reasons: (1st) the length of time before we calculate to do a business sufficiently large to make it worth while to incur the expense of so high a salary." By 1826 the company was ready to accede to Prince's salary request and hired him to come to Lowell at the rate of £1,000 a year.[19]

The news of Prince's hiring alarmed the management of Dover, as Williams indicated in a letter to Porter, who was still in England.

You will please recollect particularly that we have the Taunton Company to compete with under the superintendence of Mr. Yates and whose works will command at auction from 1 to 2 cents per yard more than others. . . . There is also the Chelmsford or Lowell Establishment who have removed Mr. [Allan] Pollock and substituted Mr Prince, an English gentleman purposefully imported and they are apparently determined to succeed. . . . We must have someone as foreman with you who is superior to Mr Yates or Mr Prince and such machinery as they know well, and must and shall accomplish the object.

Porter hired many Englishmen, but none of the stature of Prince. Porter seems to have spent his year mainly in studying textile printing firsthand. Although published descriptions of the techniques and equipment were available, Porter wanted to observe closely the various steps in machine printing and color mixing. This kind of information would help his arrangement of the interior space in the factory.[20]

Porter also visited machine shops and purchased equipment and tools. These efforts met with mixed results because of his unfamiliarity with machine technology. Some of his purchases duplicated equipment already at Dover, while others arrived badly damaged in transit. Importing machinery involved several other kinds of risks, for the exportation of certain machinery was prohibited by British law. Porter was warned that he "should be governed by circumstances in the shipment of calenders, printing machines or any other prohibited articles. If the risque of seizure is very evident, perhaps it would be better to pay the 25 per cent (although a most enormous premium) but if they can be shipped to some Southern port with little comparative risk, I think it would be best to stand our own underwriters."[21] Taking the chance was apparently worthwhile, for Porter was able to export successfully some basic equipment with which his newly hired workers were familiar.

Porter's most valuable contacts were with manufacturers, several of whom were scientists who had contributed to the chemistry of printing. In his *Memoir of Samuel Slater*, George S. White described the area Porter visited in England. "The large print-works of Lancashire are among the most interesting manufactories that can be visited. Several of the proprietors or managers are scientific men; and being also persons of large capital, they have the most perfect machinery and the best furnished laboratories." Porter visited the Primrose Mills of Clitheroe, Lancashire, owned by James Thomson and Thomas Chippendale. Thomson was a chemist who had traveled in continental Europe to study printing. One formula he passed on to Porter had been learned in Mulhouse, the French printing center.[22] Thomson was also credited with important chemical improvements of his own, including a method of substituting an alkaline solution for the common aluminous mordant

---

[19] N. Appleton to Timothy Wiggin, May 7, 1822, Appleton Papers, Massachusetts Historical Society, Boston. According to William Bagnall, "Sketches of Manufacturing Establishments in New York City and Textile Establishments in the Eastern States," ed. Victor S. Clark, 4 vols., typescript, Baker Library, Harvard Business School, 3:2172–73, the Merrimack Company considered this salary excessive and protested that it was more than the governor of Massachusetts was paid. Prince's response was "Can he print?"

[20] Williams to Porter, Aug. 17, 1826, Williams to David Sears, Oct. 21, 1826, vol. 2, DMC.

[21] Referring to "various invoices and letters," Williams listed the machinery that had arrived as a "wire rolling machine, a singeing machine, a drying machine, a slide lathe, a three-coloured printing machine, a clams machine, [and] a hydraulic press." Williams to Porter, Mar. 30, 1827, vol. 2, DMC.

[22] White, *Memoir of Samuel Slater*, p. 400. Regarding the formula, Samuel Dunster noted, "Mr. Gregoire of Mulhousen told Mr. Thomson of Clitheroe and Dr. Warwick who thinks it is known in no other place but Clitheroe in Lancashire." Workbooks of Samuel Dunster [1828–60], 11 vols., Rhode Island Historical Society, Providence, 1:57 (hereafter Dunster Workbooks).

that precipitated out under certain conditions. Later, in *The Chemistry of the Arts*, Porter gave a formula for a color he called Warwick's green and wrote: "The public is indebted for the method of preparing the alkaline solution of alumine to James Thomson Esq, a celebrated calico printer in Lancashire[,] and to Dr. T. O. Warwick, an ingenious manufacturing chemist of Manchester[,] for the method of fixing it. To both these gentlemen, the trade are under great obligations for numerous improvements in the art." Presumably the obligation was mutual, for one of the Dover workmen noted that Porter had paid Warwick fifty pounds for the formula.[23]

The Primrose works were not only a source of scientific information. At the time of Porter's visit, the majority of printers at Dover, recruited by Greenhalgh from other New England printing centers, were originally from Primrose—among them the block printers, the designer, two cutters, and at least one machine printer. As a testimonial to these Primrose workers, Williams, a hard man to please, wrote, "All appear to be sober industrious regular and steady men probably better than we should be likely to get were we to go to the expense of importing anew." Williams's fears about bringing Englishmen directly to Dover were based on his prior experiences with weavers and other workmen who have come to the company. These fears were borne out to some extent by the men Porter hired in England. Even though some of the people Porter chose had been recommended to the Dover management by former Clitheroe men, the chemist failed to foresee that inequities in the contracts he made and different working conditions in America would create unrest and disturbances that would delay productive printing at Dover beyond original expectations.[24]

The company's early commitment to block printing, as the directors were unhappy to learn, was no guarantee of immediate success. Production was hampered by a lack of workmen, inadequate supplies, and the inconvenience of sending work back and forth between the upper and lower factories before building four was ready for use. In the first full year of printing, 1826, some prints could find a market only in New York, having failed to gain acceptance in Boston. One of the first problems had been the purchase of badly worn blocks from a Boston engraver. It soon became clear that the printers had tried to compensate for the poor condition of the blocks by printing diagonally, but the auction house to which the goods were consigned protested: "No patterns are so bad as the diagonal figures. However bad the patterns may be it does not improve them by the diagonal stripes. They will not sell."[25]

As had been discovered with block printing, time was needed to bring together all the elements necessary for cylinder printing. After the company's unsuccessful experiment with Peaslee's machine, efforts at cylinder printing were given up until designers, engravers, and equipment arrived from England. The company needed the capacity for both block and cylinder printing, for one did not necessarily supersede the other. As Williams reminded the directors in his annual report for 1826, "You will recollect that cylinder work must be subsequently blocked to a certain degree, some more and some less." Early in 1827 the first cylinder machines Porter had purchased arrived in Dover. Williams, feeling pressure from the directors, who did not like delays, was anxious "to start our blue room at once as the three cylinders received from Dr. Porter were engraved for blue work and as our Goods many of them are better calculated for blue colours than any others and [we can] push them into the market as soon as possible." Shortly afterward he instructed Moses Paul to begin cylinder printing. "After this cloth is finished you will follow the cylinder entirely which will give you great dispatch compared with block work. You will, therefore, see that not a single block is laid on a piece of white goods until they have passed through the printing machine."[26] In a letter to Porter at about the same time, Williams indicated the extent to which block printing operations continued even after the introduction of cylinder work. He ex-

[23] Porter, *The Chemistry of the Arts; Being a Practical Display of the Arts and Manufactures Which Depend upon Chemical Principles*, 2 vols. (Philadelphia: Carey & Lea, 1830), 1:790; Dunster Workbooks, 1:75.

[24] Williams to Shimmin, letter 120, Dec. 4, 1826, Williams to Porter, Dec. 22, 1826, vol. 2, DMC.

[25] Bridge to Shimmin, quoting a letter from Whitwell and Bond, letter 114, Nov. 21, 1826, vol. 2, DMC. "Our printers complain Halliday blocks not having good faces. It requires a day's work to each block to perfect it. Please mention the circumstance." Williams to Shimmin, letter 143, Dec. 14, 1825, vol. 2, DMC.

[26] Report to the Board of Directors, Jan. 31, 1827, Williams to Shimmin, letter 21, Mar. 26, 1827, Williams to Paul, Apr. 19, 1827, vol. 2, DMC.

pected that "to have our establishment in full operation will undoubtedly require forty or fifty block printers." An inventory of the new printery taken in 1834 shows this to have been a remarkably accurate forecast.[27] The contents of the blocking rooms were then listed as

| | |
|---|---|
| 40 Blocking tables | $ 400 |
| 40 Benches & 30 Cases | 220 |
| 40 Frames and Rollers | 300 |
| 1 Block calender | 300 |
| | $1220 |

From the start of the enterprise, the directors had been concerned with the style and quality, hence the salability, of the work. The treasurer noted his "opinion that some of the work . . . is very imperfect"; he, being a Boston merchant like the other directors, was well aware of the nature of goods arriving there from England for sale. In one attempt at improvement Shimmin sent Williams a book with "samples [of] Taunton work patterns notched and also a few English patterns." Naturally, the directors were anxious for the Dover products to be competitive, and they made specific suggestions for the type of work to be done, keeping in close contact with Williams and Bridge at the factory. Reporting to the directors at the end of January 1827, Williams wrote that "after the completion of what is now in a state of partial preparation . . . we shall be dependent on you for the style of work to be pursued or rather the pattern we shall follow." Hardly a month had passed before Williams had to make a more urgent appeal to Whitwell, Bond and Company, the mercantile house to which Dover goods were consigned for sale. "I really wish you would send us five or six patterns for block printing which will suit Spring sale, [so] that our cutters may be kept employed."[28]

The Englishmen skilled in both block and cylinder designing who were imported to work for the American companies were a source of designs as well, although Williams did not feel that Peter Bogle, whom Porter had hired, offered any fresh ideas. "The pattern book exhibited by Mr. Bogle and the only specimen of his work he exhibits except a very ordinary shawl are not superior to our Upper Factory work, which is really bad enough," complained Williams. "I beg of you to collect all the handsome patterns you can as they may be of some future value."[29]

It was Williams's task to bring these suggestions and sources for patterns and styles together. At intervals he sent pattern samples to the directors for a decision. Four patterns were sent down after they had been "newly coloured up." Williams pointed out that "of all these we have the cylinders having received them from England." In another communication with Whitwell, Bond and Company about a pattern, Bridge indicated that he had consulted Peter Bogle about its production. "Mr. Bogle states that he can give no opinion on the expense of uniting the chocolate ground with the yellow sprig . . . [although] in England he had practised a style of work including those colors."[30] Ultimately, it was decided that it would be cheaper to print the pattern with white sprigs.

By both letters and visits, the directors guided the choice of patterns. A large assortment of prints was prepared for shipment to Boston during the spring and summer of 1827. Bridge enthusiastically reported to Shimmin in June 1827 that the most recent prints were "far superior in point of figure and in every respect to anything you have seen from here." Following the directors' suggestions, the engravers and block cutters had prepared patterns with stripes and plaids, which dominated production that summer. In addition to the sprig pattern, there was a tobacco flower motif and a pattern of white spots on a single-color background. Several cases of prints were sent from Dover in September with a pattern book illustrating the work and a note from Bridge that five patterns were "exclusively cylinder work," while three were "cylinder and block combined, the yellow and orange being put on with the block."[31]

[27] Williams to Porter, Mar. 30, 1827, vol. 2, DMC; Inventory of Real Estate, Cocheco Manufacturing Company, DMC Harvard, p. 11.

[28] Williams to Shimmin, letter 98, Oct. 16, 1826, Williams to Paul, Oct. 19, 1826, Report to the Board of Directors, Jan. 31, 1827, Bridge to Whitwell and Bond, Mar. 3, 1827, vol. 2, DMC.

[29] Williams to Porter, Apr. 3, 1827, vol. 2, DMC.

[30] Williams to Shimmin, July 28, 1827, vol. 3, DMC; Bridge to Whitwell and Bond, Aug. 10, 1827, vol. 3, DMC.

[31] Bridge to Shimmin, letter 51, June 8, 1827, vol. 2, DMC. James Lawton, an Englishman, was responsible for some of these patterns, one described as a "design . . . for 3 col'd machine and 2 blockings." Porter to Whitwell and Bond, Sept. 19, 1827, vol. 3, DMC. Porter, *Chemistry of the Arts* 2:795, prescribed a combination of block and cylinder work for a color effect of black, yellow, and orange on a blue ground.

Porter returned from Europe in August 1827 and found the printery definitely in operation, with "seventeen tables . . . and about 3,500 pieces of cloth in various stages of completion." [32] Porter also had his first experience in working with a stubborn Englishman, Peter Bogle, whom he had hired and who had been the acting superintendent of the printery in his absence. Porter immediately took charge as resident chemist, returned some poor quality chemicals to Boston, and began ordering supplies from the Manchester area. He complained to the management that "contrary to what I understand to be Mr Bogle's agreement with this company scarcely a day elapses without his pointed and unequivocal refusal to communicate with me on the subject of the processes of printing and bleaching and the loss occasioned thereby is ruinous to the concern." Porter cited three particular instances, including Bogle's refusal to work with Whitaker, the color mixer, and his disinclination to share information "as to the best method of purifying a sulphate of zinc although he expressly declared that he could do so and admitted that our sulphate of zinc requires it. The injury sustained by the impurity of this article within a few days has been very great. No less than several thousand yards of calico have been more or less injured by it." Porter received permission to dismiss Bogle, but his victory was a short-lived, hollow one. Within a week the case had a "sudden and melancholy termination": Bogle was dead of cholera.[33]

Through his experiences with Bogle and other Englishmen, many of whom were radical in their ideas about workingmen's rights and unwilling to conform to long American working hours, Porter came to agree with Williams on the need for Americans to acquire the skills to operate the printworks. Porter and Williams therefore adopted a policy of hiring American apprentices to learn from the skilled Englishmen. They pursued this policy until a new directorship decided it was impractical, and Williams and Porter were themselves let go. While the experiment of hiring American apprentices was

being conducted, Robinson, the block cutter, was chosen to supervise the apprentices, and Whitaker, the color mixer, also taught new workers. The names of two of the apprentices are known. Samuel Dunster began work at Dover in the machine shop and then moved to the printworks, where he spent fourteen months as an apprentice, from August 1828 through October 1829. In his workbooks, which have survived, Dunster indicated that Asa Alford Tufts, who later became the postmaster in Dover, was an apprentice at the same time. The workbooks, which document much of Dunster's long career as a printer, beginning under the tutelage of Robinson, Whitaker, another printer named Edmund Barnes, and Porter, are a valuable record of the principles and practices of the early American printed textile industry. The first, which contains a few samples, is a detailed record of Dunster's introduction to the "art, trade and mystery" of the calico printer. It shows how each of the Dover workmen taught his own specialty, working with the apprentice and commenting on his progress. It also reveals a great deal about the work done in the printshop at Dover.[34]

Dunster began by learning the use of the Twaddell hydrometer for determining the specific gravity of a substance.[35] Chemical manufacturing techniques were still so rudimentary that successive batches of dyestuffs often varied in strength and were sometimes contaminated by impurities. The hydrometer was used to achieve consistently correct proportions and also to test the specific gravity of shipments of chemicals received by the printery. "The two casks of acid from Cloutman stand the one at 5½ and the other 6 degrees of Twaddle," complained Bridge of a shipment of pyroligneous acid, "when the agreement was that it would have been seven degrees. Dr. Porter says also that it contains more iron than the other and of course is not so valuable." [36]

---

[32] Within a month all steps in printing were expected to be consolidated in the new factory, eliminating the cumbersome procedure of shipping cloth back and forth between the upper and lower factories in various stages of completion. [Williams], Report to Shimmin, Aug. 24, 1827, vol. 3, DMC.

[33] Porter to Williams, Sept. 18, 1827, Williams to Shimmin, letter 96, Sept. 23, 1827, vol. 3, DMC.

[34] "If the Rhode Island Historical Society think my diary of personal calicoe printing . . . worth preserving I give [it] to the Society," Dunster wrote in his will (1884), Probate Office, Bristol County Court House, Taunton, Mass. The eleven volumes of Dunster Workbooks, particularly the later ones, are filled with samples of prints documenting Dunster's career until 1860.

[35] Porter's chemical text devoted more attention to the use of the different types of hydrometer than did Gray's. Porter, *Chemistry of the Arts* 1:180–81.

[36] Bridge to Shimmin, letter 57, June 30, 1828, vol. 3, DMC. Joseph Cloutman of Salem, Massachusetts, was a man-

After Dunster had learned the principle of the hydrometer, he was taught, on September 19, 1828, how to make "simple gum water," followed by "berry yellow liquor" the next day, and the bleaching solution using chloride of lime within the week. He later learned how to combine these formulas and how to use such techniques as padding and dunging to achieve different styles of printing. Padding was a machine process for spreading mordant evenly over a piece of cloth either before printing a discharge formula on the cloth or before adding a second color. The special method of padding was required because simply dipping the cloth into a liquid mordant was insufficient. Dunging referred to the use of cow dung in the printing process. After mordanting the cloth, the printer dried and "aged" it by passing it through a solution of cow dung in water and then thoroughly washing it. Porter emphasized the "great care and judgment" needed in this process, which prevented the mordant from spreading in the dye bath and which, in Porter's opinion, enhanced the madder color.[37]

Dunster was working with much more experienced men who were themselves still trying to master many of the techniques of printing with the materials available in Dover. Dunster noted difficulties when Porter tried to use a formula given him in England by Dr. Warwick. After the first trial, Porter suggested some modifications, and the experiment was repeated with slightly better results. After Dunster had been working in the printshop for a few months, the men attempted to perfect a relatively new technique for chrome discharge. In this process an acid was used to bleach out the background color, permitting the printer to produce white figures on blue-dyed cloth or on bandanna handkerchiefs with a deep red background. Finally, it was decided that the fifth attempt, though "not perfectly satisfactory, . . . was better than heretofore." On another occasion Dunster himself tested some brown salts (the common reference for a manganese compound) sent from Boston, which appeared to be far inferior to those Porter had sent from Manchester. "Dr. Porter wished me to evaporate some in our copper boiler

and in doing it, it dissolved the kettle so as to let the liquor into the space between the kettles and it could not be effected in such a vessel."[38] Porter theorized that the composition of the imported chemical must have differed significantly from the domestic product.

From Edmund Barnes, Dunster learned about steam printing. Barnes was an English printer who had worked at Blackford Bridge, near Bury, Lancashire, and had come to Dover sometime before the end of 1829. Barnes also left a workbook containing many formulas and descriptions of steam prints —colors made fast by the use of steam (Fig. 5). This technique was in use at Dover by the summer of 1827, when large-scale printing was begun there. One of the earliest steam print designs described was on a "crimson and red ground . . . with a sprig." Barnes told Dunster that when he first came to Dover, he had had to make steam prints "before the steam cylinder was ready." The cylinder, when constructed, "stood perpendicular on one end and was enclosed in a little room like a closet and that little room was ventilated with an iron funnel like a stove pipe."[39]

From his teachers Dunster learned, often by trial and error, many details about working with colors and other chemicals. One of the indigo recipes Barnes taught him was for a color known as pencil blue. In his notes Dunster stated, "Barn[e]s said he thicken[ed] pencil blue with glue in England for the block instead of gum and it did better." Barnes may well have told Porter the same, for although the directions for pencil blue in *Chemistry of the Arts* listed gum Senegal or British gum as thickeners, Porter added that "some printers thicken pencil blue with glue, when worked by the block." Next to some of the formulas, Dunster noted his teachers' comments about them. For example, "Mr. Robinson called this [No. 10 buff] good and said there was enough gum to keep it even." Printers had to learn how to use thickeners effectively, and Dunster found that the color formula for "fast yellows," when used "for fine patterns or few stripes that should be well defined as Bengal stripe where it would be apt to run, may be thickened with 6 lb

---

ufacturer of pyroligneous acid. *Twaddle* appears to have been a local variation of *Twaddell* for, as Dunster noted, "The rule for changing the degree on Twaddle (familiarly so called . . .)," Dunster Workbooks, 1:1.

[37] Dunster Workbooks, 1:3–8, 52, 89; Porter, *Chemistry of the Arts* 2:757–58, 718–20, 723.

[38] Dunster Workbooks, 1:40–41, 90.

[39] Bridge to Shimmin, letter 56, June 16, 1827, vol. 3, DMC; Dunster Workbooks, 1:77, 80, 82. Edmund Barnes (or Barns) worked at Dover probably between 1829 and 1831. The last reference Dunster makes to Barnes's being in Dover is in May 1831.

FIG. 5. Steam green prints from Dover, Nov. 18, 1829. From Edmund Barnes, Sample Book for Printed Cottons, p. 33. (Courtesy of the Cooper-Hewitt Museum of the Smithsonian Institution, New York.)

FIG. 6. Sample of Dover print made between 1829 and 1831. From Workbooks of Samuel Dunster [1828–60], 11 vols., Rhode Island Historical Society, Providence, 1:107. (Photo, Old Sturbridge Village.)

flour or 6 lb British gum," which was actually a burned starch. Dunster noted: "Experiment proves that colour thicken[ed] with gum is preferable in point of depth and brilliancy of shade to that thickend with flour on the face side of the piece but vice versa on the wrong side." [40]

Dunster's workbooks illustrate graphically the technical problems encountered in the new printworks at Dover (Figs. 6, 7) and serve to emphasize the directors' frustration about the quality and quantity of production. While the directors were concerned with salable prints, the color workers were trying to discharge their responsibilities satisfactorily, sometimes under difficult conditions. Technical, managerial, and marketing difficulties combined to make the printworks even more experimental than anticipated. Producing perfect, marketable prints was far more complicated than simply erected a building and importing workers and machinery. When adequate machinery could be obtained, even experienced English workmen required time to adapt themselves to American conditions. Work time was lost in wage disputes and other disagreements between workers and man-

agement, as well as in waiting for needed supplies. The fact that Englishmen and Americans called the same dyestuffs by different names caused confusion in ordering. American supplies, varying from one shipment to another, were frequently of poorer quality than those to which the printers had been accustomed in England. These and other problems, along with the need to find salable designs, affected production and adversely influenced the company's reputation.

The difficulties encountered at Dover in attempting to fill an order placed by the Hamilton Company of Chelmsford for printing a large quantity of woven jean illustrate some of the problems to be surmounted. Not only was Hamilton's jean wider than Dover's calicoes, but it was twilled, so that "ordinary calendering" (the mechanical application of the final glaze and stiffening) on Dover's machines was doubly impossible.[41] Dover's costs proved to be higher than estimated, and filling of the order was hindered by serious production problems that would have occurred even if the cloth

[40] Dunster Workbooks, 1:50, 46; cf. Porter, *Chemistry of the Arts* 2:789–90.

[41] For an extended discussion of calendering, see Charles Tomlinson, ed., *Cyclopedia of Useful Arts, Mechanical and Chemical*, 2 vols. (London and New York: George Virtue & Co., 1854), 1:272–75.

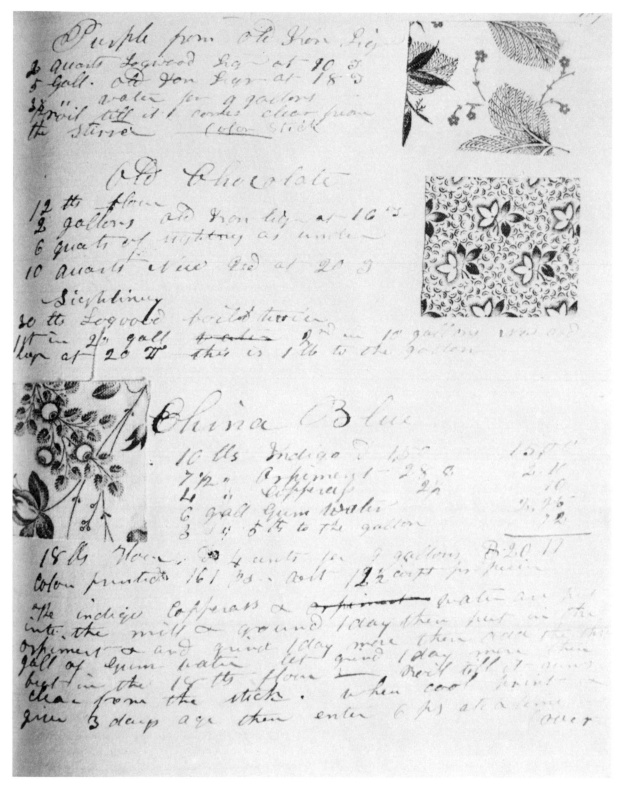

had been the correct size for the Dover cylinders. The description of the imperfections and their causes that accompanied the cloth when it was eventually delivered indicates the accidents and lack of scientific procedure that plagued the establishment of a textile industry in America. "Those pieces which are deeply stained crosswise of the cloth so as to involve a great part of the piece, we attribute to stains in the grey cloth which could not be removed by bleaching and did not appear until the cloth was dyed up." Several pieces of the cloth had been "injured by rats in the drying room." But the greatest difficulty stemmed from "steam condensing and dropping from the roof of our dyehouse, the slate of which is fastened with iron instead of copper nails. . . . [The] imperfections consist of dark and generally small stains." The printers had been able to improve the last-named condition by discharging or chemically removing the color from the cloth; but, in order to avoid any residual chemical action from the original printing, the cloth was "printed in buff the second time lest a trace of mordant might remain." It was perhaps a consolation for the Hamilton Company that the shipment of finished goods represented an "increase of 1019 yards . . . owing of course to the effects of calendering." The goods were not warranted as first quality. "It was never expected," wrote Williams, and apparently other printworks were equally unable to cope with the problem of quality control. "Mr. Prince informed me that Lowell did not warrant the Jeans which they [had] done perfect nor could they for the price charged." [42]

As this example shows, in the 1820s the production of large quantities of printed cloth involved the meshing of more variables than even an adequately financed and professionally staffed company like Dover could fully control. Expenses had been greater than anticipated, and from the beginning Shimmin had been concerned about delays. Progress was slow, costs were high, and the wait for production and for the income from sales seemed endless. The president of the company, David Sears, visited Dover in 1826 and inspected the factories, having been unable to obtain firm cost estimates for the new printery and auxiliary buildings.

Although his visit came several months after the legal expansion of the company's capital,[43] it must have been evident by then that the printing operation, with final completion still in the distant future, was draining the company's resources. The next year it was decided to seek a separate act of incorporation for the printing operation, and the Cocheco Manufacturing Company was chartered, with an authorized capital of one and a half million dollars. Williams, who was also a member of the New Hampshire legislature, felt that this move created some suspicion among his colleagues in Concord.

You will notice that I have omitted to insert Mr. Sears, Mr. Shimmin[s] and many other names who are proprietors and have inserted other names, some of whom are not proprietors. Whether Mr. Jackson is a proprietor in this stock or not I did not feel certain, but knew he was a stockholder in the D. M. Co., therefore took the liberty to incorporate his name in the Act, as I found an impression at Concord that this was a money Act and it was with considerable difficulty that I could convince some "knowing ones" that we seriously wished the act for a manufacturing company.[44]

The strategy worked and the charter was granted, although it led to no immediate changes in the way affairs were managed. Williams, as agent of the Dover Manufacturing Company, continued to oversee the operation of the Cocheco printing establishment.

Changes in administration were precipitated, however, by financial difficulties in May 1828 which threatened the stability of banks in both Dover and Portsmouth. Williams resigned as agent and director of the Dover bank, and for the next several months Matthew Bridge was the agent of the Dover Manufacturing Company. The Cocheco Manufacturing Company finally received a formal transfer of the printworks in an auction at the end of 1829, at which David Sears bid a dollar more than the debts of the Dover Manufacturing Company, most of which were owed to Cocheco. By then, a separate management had been initiated at Cocheco under James F. Curtis, who remained as agent until 1834, when he was replaced by Moses

[42] Williams to William Appleton, Nov. 26, 1827, Feb. 23, 1828, Williams to Shimmin, letter 16, Mar. 16, 1828, vol. 3, DMC.

[43] New Hampshire, House of Representatives, *Journal*, June 20, 1826, p. 137.
[44] Williams to Shimmin, letter 63, July 2, 1827, vol. 3, DMC.

Paul, the former superintendent of the upper factory.[45]

The Cocheco Manufacturing Company took over a textile factory and a printworks ready for production and proceeded to restaff the latter in an effort to guarantee that its output would become profitable. Among those who lost their positions during these months as part of the new policy were the mechanics Jonathan Fiske and John Chase, and Arthur L. Porter, Asa A. Tufts, and Samuel Dunster from the printery. Many years later Dunster recalled that Curtis had told him, "I was driven to part with you for the new folks [would] not have an American on the [print] works." [46] Such a policy represented a direct reversal of John Williams's reluctance to depend completely on English workers, feeling that they caused unnecessary trouble and obstructed work. Williams felt his policy had been borne out in the negotiations he had had with English printers, who had not expected to work long American hours, objecting on the grounds that color work required better light than was available in the early and late hours through which weavers and spinners worked. The reorganization, while it caused personal hardships for the individuals who lost their jobs, helped assure that the printworks at Dover would become a viable commercial operation. The decision to depend on British labor emphasized the continued reliance of American printworks on British skills. Meanwhile, Porter and Dunster went their separate ways. Porter remained in Dover as a physician until 1836, when he moved to Michigan. Shortly after leaving the company, he arranged with a Philadelphia publisher to prepare an American edition of Gray's *Operative Chemist,* which had been published in England in 1828. As the first American chemist to work in a calico printing factory, Porter was able to expand Gray's work to include information learned through his own experiences with the English calico printers in Lancashire and at the Dover Manufacturing Company. In his book he combined chemical knowledge with a good business sense, emphasizing cylinder printing technology and large-scale factory production

methods, and thus making technical information available to an expanding industry.[47]

After his release from Dover, Dunster worked as a printer in Pennsylvania, Maine, and Rhode Island. He returned to the Cocheco printworks as the assistant superintendent in 1852, at which time Moses Paul was the company's agent. Dunster remained at Dover for six years and retired from printing shortly afterward.[48] Throughout his career Dunster saved samples of his own work, as well as those of the work of some of his contemporaries. The prints dating from the years he worked in Johnston, Rhode Island (1844–47), are worthy of special attention because they are unlike any others recorded in his books in fabric, design, and color. The coarse cotton was printed with striking geometric designs in bold colors, perhaps fitting the description that Whitworth and Wallis gave, in their report to the House of Commons, of cloths made for the southern and western markets.[49] In contrast, the samples of work from the Cocheco printworks (1852–58), as well as those from the Allen companies, for which Dunster worked in Rhode Island, are calicoes printed in madder colors, purples, pinks, and reds. Unlike Dunster's patterns from Johnston, Rhode Island, these calicoes were greatly influenced by British designs and colors, for which Dunster expressed his admiration and which he clearly attempted to reproduce.

The earliest products of American printworks have fallen into a disfavor from which they have not yet emerged, despite John Williams's assertion that he was "fully aware of the importance of suiting the public taste." Yet even by midcentury, American prints were judged by knowledgeable contemporaries as very similar to the work of British manufacturers (Fig. 8). Whitworth and Wallis

---

[47] Another chemist who later played an important role in textile printing was Samuel L. Dana. Porter's career paralleled Dana's. Both men received medical degrees in 1818. Dana practiced medicine in Waltham before establishing a plant there in 1826, where he manufactured sulphuric acid and bleaching materials. Following a study visit to Europe in 1833, he became the chemist for the Merrimack Manufacturing Company.

[48] Edwin Emery, *The History of Sanford, Maine 1661–1900,* comp. William M. Emery (Fall River, Mass.: by the compiler, 1901), p. 216; Dunster Workbooks, vols. 2–11; Samuel Dunster, *Henry Dunster and His Descendants* (Central Falls, R.I.: E. L. Freeman & Co., 1876), pp. 312–14.

[49] Dunster Workbooks, vol. 4; Rosenberg, *American System of Manufactures,* p. 245.

---

[45] "The Old Nail Factory," Historical Memoranda 3, no. 391, Dover Public Library; John Scales, "The Beginning of Our Industry in Dover," *Cocheco Chats* 1 (Oct. 1821): 7, Dover Public Library.

[46] Dunster to ?, June 28, 1880, Historical Memoranda 3, no. 394, Dover Public Library.

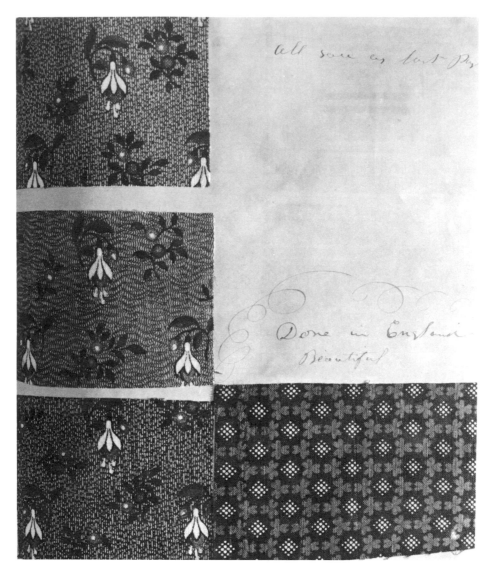

FIG. 8. On the left are three madder prints done at Cocheco Manufacturing Company, ca. 1852. From Workbooks of Samuel Dunster [1828–60], 11 vols., Rhode Island Historical Society, Providence, 9:131. (Photo, Old Sturbridge Village.)

wrote of American prints that "the patterns are generally well selected, being of a neat character and similar in many respects to the goods known as 'Hoyles.' The smaller kind of children's wear, are especially noticeable for the same qualities which under the prints of such [English] houses as Hoyle's and Liddiard's [are] so much in request in England." The influence of Hoyle's work was felt in the United States in ways that can be documented. Merrimack's John D. Prince had been apprenticed to Thomas Hoyle at around the turn of the nineteenth century. While at Cocheco in the 1850s, Dunster made several attempts to duplicate a sample of cloth printed in a small red design on

pink, stamped "Thomas Hoyle and Sons, Manchester, British cambric."[50]

As the experience at Dover and other printworks shows, the American textile printing industry during this period was largely dependent upon skilled Englishmen who were brought to the United States. Considering the source of the chemical, mechanical, and design technologies, it is hardly surprising that it remains very difficult to determine whether early Victorian printed cottons were of American or British manufacture.

[50] Rosenberg, *American System of Manufactures,* p. 245; Bagnall, "Sketches of Manufacturing Establishments," 3: 2172; Dunster Workbooks, 9:140–41.

# Design Sources for
# Nineteenth-Century Window Hangings

*Samuel J. Dornsife*

IN ALMOST ANY RESTORATION or re-creation of a nineteenth-century house or period room the problem of appropriate window treatment arises sooner or later. The importance of window treatment in a decorative setting is indisputable, yet no one museum or library has a large or comprehensive collection of material focusing on changing styles of drapery arrangement. While quantities of nineteenth-century drapery designs have survived, the curator's search for them is often long and arduous. As an aid to restoration processes and as a hint of the richness of the material available in the field, this pictorial essay presents a selection of representative nineteenth-century window dressings from the most important European and American printed sources.[1]

Drapery designs throughout the nineteenth century reflected the capriciousness of fashion in the homes of the wealthy and the succession of revival styles that gained popularity in domestic furnishings. Although surviving evidence of eighteenth-century drapery treatment is meager, by the mid-nineteenth century arbiters of taste recommended curtains for "comfort and elegance" even in modest homes and emphasized the necessity of "taste and judgment" in selecting drapery styles. Charles Percier's and Pierre Fontaine's 1827 description of the reign of fashion in French circles applies equally to high-style society in other countries:

The condition and custom of modern societies, which put all individuals on exhibition wherever they walk, talk, gamble and play, have aroused to the highest degree the desire to please on the one hand and the wish to be noticed on the other. That explains the dominion of Fashion in everything that concerns dress, finery and manners; it also explains that everlastingly renascent impulse that leads the many to imitate the few who set the style and which prompts the few to abandon it as soon as it becomes wide-spread. The more the taste and pleasures of what is now referred to as *Society* have increased, the farther the impulse of Fashion has extended its power, and there is almost nothing in the interior of homes that is not subject to it. Decoration and furniture become for homes what clothes are to people: everything of this sort ages, too, and in just a few years is considered old-fashioned and ridiculous.[2]

It is hardly surprising that in this milieu of rapidly changing taste, many drapery designs, or collections of designs, were never published as shelf books. More usually they were issued as single plates or as groups of plates intended as working tools for the professional or amateur draper, who would use them until they were tattered or out of fashion and then would discard them.[3]

[1] All illustrations are from the Dornsife Collection, which is now the property of The Victorian Society in America at the Athenaeum of Philadelphia, although some books have yet to be physically transferred to that institution.

[2] Anna Brightman, "Window Curtains in Colonial Boston and Salem," *Antiques* 86, no. 2 (Aug. 1964): 184–87; Thomas Webster, *An Encyclopaedia of Domestic Economy* (New York: Harper & Bros., 1845), p. 249. Percier and Fontaine, *Recueil de décorations intérieures* (1827), quoted in Paul Cornu, Preface to A. Calavas, ed., *Meubles et objets de goût 1796–1830: 678 documents tirés des journaux de modes et de la "collection" de La Mésangère* (Paris: Libraire des Arts Décoratifs, [1914]), pp. 1–2.

[3] No comprehensive bibliography of drapery design sources will be attempted here since adequate published bibliographies exist. For English materials, see Jeremy Cooper, "Victorian Furniture: An Introduction to the Sources," *Apollo*, n.s. 95, no. 120 (Feb. 1972): 115–22. For French materials, see Jacqueline Viaux, comp., *Catalogue matières: Arts-décoratifs, beaux-arts, intérieures*, 4 vols. to date (Paris: Société des Amis de la Bibliothèque Forney, 1970–). Clive Wainwright of the Department of Furniture and Woodwork, Victoria and Albert Museum, is also preparing a comprehensive bibliography.

The two most important sources for window arrangements suitable to the early part of the century are collections published by Pierre de La Mésangère and Rudolph Ackermann, respectively. La Mésangère (1761–1831), who was at one time a candidate for the priesthood, founded, with an associate named Sellèque, the *Journal des dames et des modes* in 1797. In 1799 he became sole editor of the magazine, whose ladies' fashion plates included details of furniture and decorations. In 1802 La Mésangère announced his intention of publishing a separate series of plates on furniture and decorative objects:

> A precise nomenclature for objects of taste and fancy executed by our cabinet-makers, engravers, jewelers, and especially our gold and silver-smiths would be neither less curious nor less extensive than that found in English magazines; the difficulty of making these objects recognizable without the help of the picture has prevented us, until now, from speaking about them; but we already have five engraved furniture plates; two plates of clocks will appear before the end of the month, and the designs for a dozen plates of gold and silver objects and jewelry are in the hands of the engravers. All of the plates are in color, and since they are four times larger than those of our costumes, we could not, without damaging them by folding them many times, send them through the mail; our plan is to combine them in albums of ten sheets; each album picked up at our office will cost three francs.[4]

The plates that formed *Meubles et objets de goût* appeared irregularly from 1802 to 1835. About 100 were published in 1802 and 1803, 150 between 1804 and 1807, another 150 from 1808 to 1814, and later about 18 plates a year. Those published after La Mésangère's death, in 1831, carry the caption, "Couet, auteur-éditeur rue de Vaugirard, no 15," in addition to the serial number and "Paris," which appeared on all the plates. In all, 755 plates were issued; although scattered groups of plates can be found in several locations, to the best of the author's knowledge no complete set of the 755 plates has survived.[5]

Rudolph Ackermann (1764–1834), who left Saxony and settled in England in the 1780s, was the publisher of a journal of fashion and taste, *The Repository of Arts, Literature, Commerce, Manufactures, Fashions, and Politics* (1809–28). This magazine, as well as many of Ackermann's other publications, reflects the publisher's interest in promoting what today would be called interior design and carries numerous illustrations of fashionable settings and decorative objects. Files of *The Repository of Arts* occur with some frequency, since Ackermann apparently sold both current and back issues and offered bound volumes of back issues to buyers who wanted to complete their sets.[6]

In the early years of the century, fashion dictated a simple, spare, almost austere "Pompeian" style. In France, as the Empire waxed, decorative motifs shifted from a strictly classical vocabulary to details more indicative of the cult of Napoleon: palms, crowns of laurel, swans, and bees, for example. La Mésangère's designs reflect a light, casual, bourgeois interpretation of the prevailing Empire style (Figs. 1, 2). Percier and Fontaine, Napoleon's official arbiters of taste, complained that La Mésangère's interpretations rendered their designs so economical as to be within the reach of the smallest fortune. But Ackermann cautioned his readers that the "difficult and important branch of the upholstery art, drapery in general, requires the talents of the draughtsman, combined with professional experience and taste" (Figs. 3, 4).[7]

Three other Frenchmen, known only by their surnames, Hallavant, Osmont, and Pinsonnière, published collections of drapery designs in the early decades of the nineteenth century. Each of them published albums of fifty plates, costing twenty francs plain and twenty-five francs in color. Other than what survives of these albums, the careers of these men are unknown. These three collections, as well as the works of La Mésangère and Ackermann, indicate that well into the 1830s designs were based primarily on classical examples, although more and more complexity was introduced into window hangings. Designs required more material, more trimmings (fringes, gimps, galloons, ropes, and tassels), and more elaborate hardware (poles, pelmets, and tiebacks), while drapers attempted more frenzied convolutions and combinations. At the same time there developed a decided aversion to treating two windows in the same wall as separate units, and fabric swirling, often in several layers, and combinations of poles

[4] La Mésangère, *Journal des dames et des modes* (Mar. 16, 1802), quoted in Calavas, *Meubles et objets de goût*, p. 9.
[5] Calavas, *Meubles et objets de goût*, pp. 9–10.

[6] W. J. Burke, *Rudolph Ackermann: Promoter of the Arts and Sciences* (New York: New York Public Library, 1935), pp. 3–33.
[7] *A Series, Containing Forty-four Engravings in Colours, of Fashionable Furniture* (London: R. Ackermann, 1823), p. 2.

or pelmets were used to link separate windows (Figs. 5–7).

Interpretation of early drapery designs, including identification of the materials used, is difficult because most were published without comment. A few have brief notes attached—for example, La Mésangère occasionally mentioned taffeta, mousseline, and bazin. An occasional plate has fuller marginal notes, apparently written by a draper, indicating yardages of materials and trimmings and the type of room suitable for the drapery style (Figs. 8–10). Despite the lack of description, drapery illustrations often include pieces of furniture and small decorative items that serve to provide some context for the usage of window displays. One of La Mésangère's stated purposes was to show the simultaneous variations of fashions ranging from clothes to carriages. And, not surprisingly, some valuable drapery illustrations appear in cabinetmakers handbooks (Figs. 11, 12). George Smith's *The Cabinet-Maker and Upholsterer's Guide* (1826) includes an unusually complete description of his designs for drawing room curtains, along with a practical critique of the impact of different drapery styles and colors.

The curtains with their draperies are supposed to be made up, either in plain sattin or damask, of which there are two kinds; the one being composed of silk altogether; the other being a mixture of silk and worsted; which last, though it may happen to be cheaper than the other, it will when cleaned or dyed, shrink considerably more. . . . Drapery will ever give consequence to an apartment, and although it may for a time be in disuse from the caprice of fashion, it will always be adopted wherever a good taste prevails; economy may render the plain valance necessary, but it never can be introduced with a view of producing a better effect; and withal when the brass rods and large rings, &c. are added, the savings becomes very doubtful. In almost every design . . . there arises a necessity for using a variety of colours, inasmuch as a gay and more lively effect is produced by the contrast; but if we refer to a more chaste style of colouring . . . it will be found in the use of one colour alone, such tint predominating throughout the whole; two other tints may be used, but the three must be of one stock, each varying from the other only by a darker or lighter gradation of the same tint.[8]

By the late 1830s at least one series of drapery designs, that published in Paris by Osmont and his successor, Pinsonnière, indicated a change from increasingly overblown classical styles to romanticism

and a revival of eighteenth-century styles. Osmont's series five and series six, undated but probably published in the mid-1830s, show only heavy-handed classical treatments (Figs. 13–15); but Pinsonnière's series seven, issued in 1839, although retaining some classic examples, shows many designs reminiscent of Louis XIV and Louis XV styles and includes designs suitable for chintz, as well as for the silk and damask continuously advocated by decorators (Figs. 16–20).

At the same time that Osmont and Pinsonnière were publishing their romanticized yet ornate designs, J. C. Loudon's *Encyclopaedia of Cottage, Farm, and Villa Architecture and Furniture,* first published in 1833 and reissued in numerous editions, testified to the popularity of draperies by showing less wealthy householders how to arrange and install simplified styles (Fig. 21). Loudon discussed window treatments for cottages and villas, and included illustrations of hardware and instructions for its installation. Arguing that windows in English cottages should be larger so curtains could be accommodated in every home, Loudon remarked on the more general use of curtains by the French. "From Stockholm to Naples, the room of a Frenchman may always be known, before entering it, by the curtains of his window." But another authority on simple interior decoration indicated that French styles were too intricate for the average home. "One inconvenience in the elegant French draperies was the great skill and taste required to put them up well." [9] Given French interest in window decoration, it is not surprising that another variety of window treatment, fabrics arranged over painted shades, was shown in a French publication. Désiré Guilmard's series of 260 color folio plates dating from 1844 to 1855, *Le Garde-meuble, ancien et moderne,*[10] includes many designs using shades (an item often neglected in restoring nineteenth-century houses), as well as arrangements of banners of matching material hung beside draperies, a treatment popular from the 1830s to the 1880s (Figs. 22–25).

America, too, did not lack publicizers of interior

[8] Smith, *The Cabinet-Maker and Upholsterer's Guide* (London: Jones & Co., 1826), pp. 175–76.

[9] Loudon, *An Encyclopaedia of Cottage, Farm, and Villa Architecture and Furniture* (1833; new ed., London: Longman, Orme, Brown, Green & Longmans, 1839), p. 338; Webster, *Encyclopaedia of Domestic Economy,* p. 251.

[10] Although the Bibliothèque Forney bibliography of French books pertaining to the decorative arts lists 260 plates in Guilmard's series, the Dornsife collection includes plate no. 665.

designs, and *Godey's Lady's Book,* published from 1830 to 1898 by Louis A. Godey, brought accepted drapery styles to the attention of the American public. At irregular intervals between 1851 and 1860, Godey printed drapery designs, sometimes with descriptive notes and prices, as advertisements for the W. H. Carryl firm in Philadelphia. Often, however, Godey's designs were repetitive, either of arrangements printed in European sources or of hangings already published in the *Lady's Book* (Fig. 26). The willingness to accept styles first introduced in the 1830s and extensively republished through the 1860s indicates the basic conservatism of Godey's middle-class Victorian readers.[11]

Midcentury publications commented more fully on designs than did earlier works, and Godey's remarks on specific designs are valuable for indicating the ways that draperies were used in different rooms, as well as the materials and colors that were preferred. According to the *Lady's Book* of 1851, usage of draperies was widespread. "Draperies are used for beds, windows and mirrors. The materials and general style of all three are the same." *Satin de laine,* damask, velvet, and brocatelle were considered appropriate fabrics, and garnet and crimson were thought to constitute "the richest combination of colors," although blue and gold, gold and green, purple and gold, blue and fawn, and black and white with crimson were also acceptable. In 1854 crimson was still a favorite color in combination with maroon, maroon and gold, gold, and white. Gold, too, was a popular combining color with various shades of green, blue, and purple. Blue and green were used together, and Godey mentioned both a "delicate rose color" and the "richest shade rose color." In addition to draperies, window shades were in constant demand. The list of popular materials expanded, too, with specific materials "approved" for different rooms: for parlors, French satin damask, French brocatelle, and French and English terry; for dining rooms, libraries, and parlors, French *satin de laine;* for plainer rooms, including sleeping quarters, librar-

ies, and dining rooms, worsted rep, printed lasting, worsted damask, and damask laine.[12]

Along with detailed descriptions of both draperies and their decorative settings, Godey indicated the seasonal appropriateness of different arrangements. His comment on figure 26, showing hangings suitable for winter, illustrates some of the rigidity of mid-Victorian ideas of taste. In this case the more typically late Victorian eclectic admixture of styles was definitely denigrated.

It will be seen that it presents the entire front of a drawing-room, the pier-glass of rich plate and exquisite setting, the window draperies on each side finished by heavy cornices corresponding to that of the mirror, and in fact forming one heavy mass of rich carving and gilding, which would do no discredit to the days of the monarch for whom it is called, Louis XIV. The curtains themselves are of crimson brocatelle, lined and interlined with white silk; undercurtain of a heavy French embroidery; all the trimmings, which are arranged with peculiar taste, are manufactured of the best materials expressly to correspond; and the marble console beneath the mirror has a rich lambrequin in the same style. Even the very chairs, and their arrangement just inside the curtain, are of the favorite style of the day, and are suited to the fashion of the cornices. We could scarcely give a more correct idea of the interior of the house-palaces of our "merchant princes."

This is a winter arrangement, the heavy folds of the silk drapery being usually taken down at the approach of a warmer season.

For warmer seasons (Fig. 27) Godey described

a summer drapery, the cornice of a lighter style, the long curtains of delicate French lace embroidery, and the lambrequin only, with its heavy garniture of fringe, cords, tassels, and gimp (all corresponding in hue and style), being of brocatelle. This is especially suited to country houses, used chiefly in the summer season, and usually more lightly furnished than a town residence. It is also a tasteful drapery for the long windows opening into the third room or saloon from the parlors, in houses arranged in that manner, being generally adopted in New York, where the best houses are built in that manner.[13]

The *Lady's Book* also abounds in illustrations of the varied revival styles popular after the decline of classical influences. In fact, so many types of window hangings were available that the *Lady's Book*

[11] *Godey's Lady's Book* 42 (Oct. 1851): 243, fig. 1, shows Roman striped draperies similar to those published by Guilmard (see Fig. 25); Cooper, "Victorian Furniture," p. 116. See also "Draperies, Curtains, and Blinds," *Godey's Lady's Book* 60 (Feb.–June 1860): 185, 282, 325–28, 505, a four-part article discussing details of designs and methods of hanging; the styles recommended were first shown at least thirty years before.

[12] "Window Curtains," *Godey's Lady's Book* 42 (Oct. 1851): 243–44; "Parlor Draperies: New Designs," *Godey's Lady's Book* 48 (Aug. 1854): 170–71.

[13] "Parlor Draperies: New Designs," p. 171.

found it difficult to estimate the costs of draperies. "The sizes and styles of curtains are so varied that no set price can be fixed; the prices vary between fifteen to two hundred and fifty dollars, according to the materials used." For Renaissance-style draperies illustrated in 1859, prices depended on fabric selection, and, as always, the reader was reminded of the importance of the combined effect of the window with its draperies and the surrounding furniture and decorations (Fig. 28).

The right-hand window represents a curtain in the renaissance or antique style. In addition to the long satin and lace curtains there are draperies of the same satin material as the long curtains, with rich silk fringe, tassels, etc. This curtain in rich material—brocatelle and lace curtains—will cost, probably, from $140 to $175 per window. The price can be graduated from $75 upwards, according to the fabric used.

The left-hand window is less elaborate, and is the style now mostly in use, when made up in rich material. Brocatelle and lace curtains have a very rich effect, and will generally please the most fastidious; the lace curtains should invariably be hung next the glass. This style in rich brocatelle and corresponding lace curtains will cost from $125 to $140 per window; in silk terry, about $10 per window less.

The colors mostly used are solid crimson, or crimson and maroon, either of which are in good taste, and have a rich and pleasing effect. The colors, however, are simply a matter of taste, and of course depend on the purchaser. In the pier below the mirror is introduced beneath the marble slab a Lambrequin of the same material as the curtains, which adds much to the beauty of the room.[14]

After midcentury even more revival styles gained prominence in Europe and America. The Empress Eugénie's fascination with Marie Antoinette created renewed interest in late-eighteenth-century fashions, although French leadership in style in the second half of the century was successfully challenged by English designers in many aspects of the decorative arts. Throughout the 1860s and 1870s new window treatments and old usages, both applying recognized styles, were found on both sides of the Atlantic. In Paris, Antonio Sanguinetti's *L'Ameublement au XIX siècle* mirrored fashion interests in France (Figs. 29, 30), while J. Wayland Kimball's *Book of Designs: Furniture and Drapery* performed the same function in America. Kimball's illustrations indicate the strong emphasis

placed on drapery settings, especially lambrequins, and some of his designs show the influence of the neo-grec style, a popular competitor of eighteenth-century revivals (Figs. 31, 32). Although design books continued to repeat previously published illustrations, Kimball included an "original design for the special decoration of a window, its main feature being the frames with shelves for plants, and the *jardiniere* with its moveable support below. The valance is introduced to add to the finish; in a very strongly lighted window a lace curtain might also be employed" (Fig. 33).

Kimball's concern with details extended to providing price information for fashionable trimmings (Fig. 34). For example, his price list indicates that drapery cords cost from $.07 to $.43 per yard, the higher price being for silk; that 6-inch silk fringe cost $1.84 per yard; and that drapery tassels varied from $6.20 per dozen for 6¾-inch silk and worsted, to $4.40 each for 11-inch pure silk (Fig. 35). Another commentator on interior design provided prices per yard for materials: cotton "momie-cloth," 50 inches wide, $1.10; wool "momie-cloth," $3.00; felting, 2 yards wide, $1.50; imported English Bolton sheeting, $1.00; stamped velveteen, $1.25 to $2.00; and double-width cotton flannel, $.90.[15] Other important materials late in the century were plush and sateen, alone or in combination, and cretonne. The perennial favorite, woolen rep, was still in common use and was frequently recommended for its durability. Embroidered designs and borders were quite popular in the 1880s.

Stylishly rendered portieres, in vogue in the 1880s, could be executed in some of the most fashionable materials (Fig. 36). "Velveteen is a desireable material for portieres or curtains. Plush is the richest material in use." Even ingrain carpet was recommended for portieres. It could be used in a number of colors, but olive and crimson were preferred because they faded "handsomely." In addition the carpet could be decorated. "Down one side of each breadth can be worked in Germantown wool, a pattern adopted from a Turkish rug." In fact, authentic Turkish rugs were "brought home by travelers from the East and imported in great quantities" for use in portieres, and "prayer

[14] "Carryl's Curtain Establishment," *Godey's Lady's Book* 56 (Aug. 1858): 171; "Curtains for Parlor or Drawing-room Windows," *Godey's Lady's Book* 58 (Sept. 1859): 265.

[15] Kimball, *Book of Designs: Furniture and Drapery* (Boston: by the author, 1876), p. 16; Almon C. Varney, *Our Homes and Their Adornments* (Detroit: J. C. Chilton & Co., 1882), p. 263.

74

carpets" were hung on walls and doors." [16] This craze for Turkish designs is typical of the eclectic taste of the late nineteenth century (Fig. 37), dur-

ing which there were at least two such periods of Turkomania and exuberance for Japanese designs. At the same time, designs of Gothic, Elizabethan, and Jacobean derivation continued to appear and disappear (Figs. 38–41).

[16] Varney, *Our Homes,* pp. 262, 263.

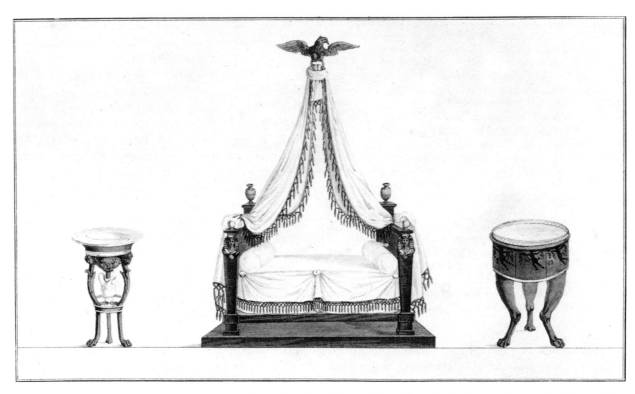

FIG. 1. Lit à la romaine. From Pierre de la Mésangère, *Meubles et objets de goût* (Paris: Au Bureau de Journal des Dames, 1802), pl. 26. Both hangings and bedspread appear to be sheer white muslin. (Photo, Winterthur.)

*Fig. 2. Draperies de croisées.* From Pierre de la Mésangère, *Meubles et objets de goût* (Paris: Au Bureau de Journal des Dames, 1802), pl. 253. *Left,* periwinkle blue and lemon yellow over sheer white curtains; *center,* orange buff over sheer white; *right,* buff and purple both edged with black. (Photo, Winterthur.)

FIG. 3. Sofa bed. From *The Repository of Arts,* 1st ser., vol. 1, no. 5 (May 1809), pl. 21. The bed drapery is in cerulean blue and gold. (Photo, Winterthur.)

**FIG. 4.** French window curtain. From *The Repository of Arts,* 1st ser., vol. 2, no. 10 (Oct. 1809), pl. 26. A color scheme of orange buff and pale blue. (Photo, Winterthur.)

FIG. 5. *Draperies jumelles*. From Pierre de la Mésangère, *Meubles et objets de goût* (Paris: Au Bureau de Journal des Dames, 1809), pl. 304. Shown in grass green and lemon yellow, an early example of tying windows together with draperies. (Photo, Winterthur.)

FIG. 6. *Draperie de croisée*. From Pierre de la Mésangère, *Meubles et objets de goût* (Paris: Au Bureau de Journal des Dames, 1811), pl. 333. *Left*, coral silk; *right*, buff trimmed in terra-cotta. (Photo, Winterthur.)

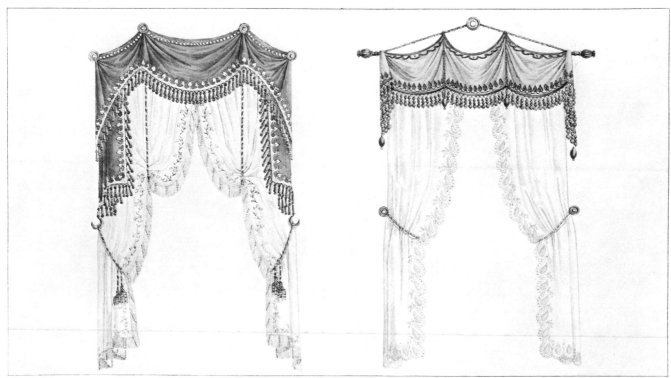

Fig. 7. *Croisées Drapées.* From Pierre de la Mésangère, *Meubles et objets de goût* (Paris: Au Bureau de Journal des Dames, 1815), pl. 406. Purple trimming on gold, with the hardware finished to resemble verdigris. (Photo, Winterthur.)

Fig. 8. *Deux Croisées.* From Hallavant, "Recueil des draperies," ca. 1815, pl. 29. An arrangement in red and gold, over sheer white curtains, spanning two windows and their pier. The pole appears to be walnut with ormolu rings and mounts. Note the draper's notes at the upper right. (Photo, Winterthur.)

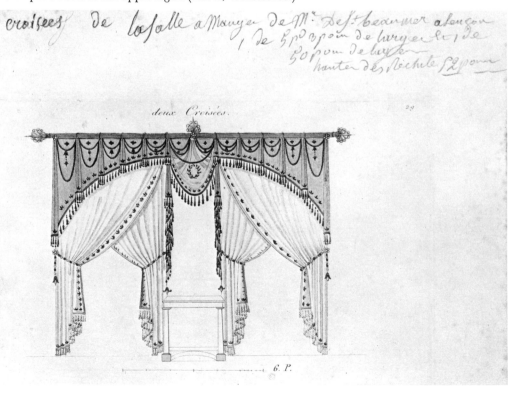

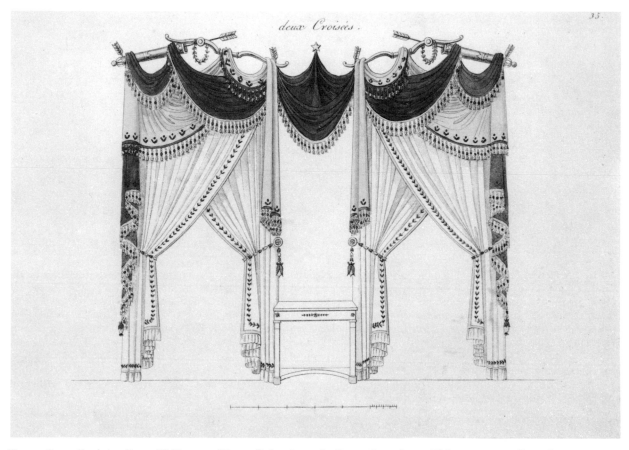

FIG. 9. *Deux Croisées.* From Hallavant, "Recueil des draperies," ca. 1815, pl. 35. This more complicated arrangement is also in red and gold over sheer white curtains, but here gilded bows, quivers, and arrows support the swags. (Photo, Winterthur.)

FIG. 10. "A Design for a Dining-Room Suite, Curtains excepted, in order to shew more particularly the lines of the Window-Architrave and French Casement. The Casement of the cornice may be of mat gold, or covered with black velvet; in the center and on the ends of which are pine-apples, with their natural leafing; clusters of grapes and leaves, carved, are formed on a strong wire (previously interwoven), to entwine the cornice." From *A Series, Containing Forty-four Engravings in Colours, of Fashionable Furniture* (London: R. Ackermann, 1823), pl. 15. (Photo, Winterthur.)

FIG. 11. Drawing room window curtains. From George Smith, *The Cabinet-Maker and Upholsterer's Guide* (London: Jones & Co., 1826), pl. 2. *Left,* light purple and medium green, both trimmed in gold; *right,* cherry red with blue overdrapery and facings. (Photo, Winterthur.)

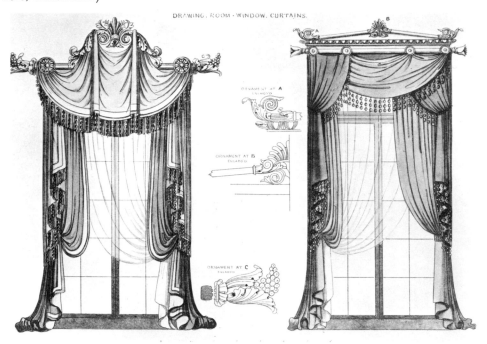

FIG. 12. French bed. From Peter Nicholson and Michael Angelo Nicholson, *The Practical Cabinet-Maker, Upholsterer and Complete Decorator* (London: H. Fisher, Son & P. Jackson, 1826), pl. 3. All sheer white fabric. La Mésangère shows the same design; its origin is unknown. (Photo, Winterthur.)

FIG. 13. *Lits jumeaux ou façade de croisées.* From Osmont, *Cinquième Cahier* (Paris: Osmont, Editeur de Dessins, n.d.), pl. 16. An early display of twin beds, with both draperies and bedspreads in gold piped with red. (Photo, Winterthur.)

FIG. 14. *Alcôve avec cabinets.* From Osmont, *Cinquième Cahier* (Paris: Osmont, Editeur de Dessins, n.d.), pl. 21. The drapery is gold trimmed with red laurel leaves and blue galloon; the blue overdraperies and bedspread are trimmed identically. (Photo, Winterthur.)

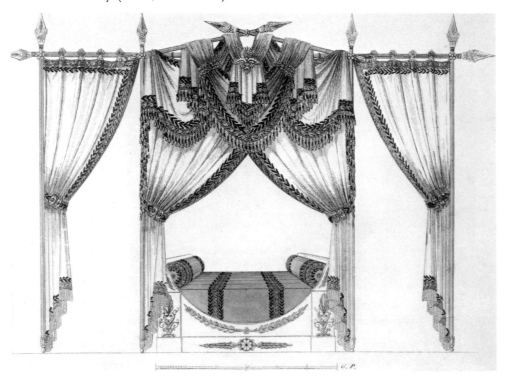

FIG. 15. *Deux Croisées.* From Osmont, *Cinquième Cahier* (Paris: Osmont, Editeur de Dessins, n.d.), pl. 44. Blue with red tassel fringe over embroidered and plain white sheers. (Photo, Winterthur.)

FIG. 16. *Croisées à bonnes grâces et câblées.* From Pinsonnière, ed., *Nouveau Recueil de draperies arrangé et litho graphié par F. Julienne* (Paris: Pinsonnière, Successor of Osmont, 1839), pl. 16. *Left,* pink draperies, probably satin, with blue ropes and tassels over lace; *right,* chintz, again with blue cords and tassels; both arrangements suspended from gilt poles. (Photo, Winterthur.)

FIG. 17. *Lit à chassis, angle coupé.* From Pinsonnière, ed., *Nouveau Recueil de draperies arrangé et lithographié par E. Julienne* (Paris: Pinsonnière, Successor of Osmont, 1839), pl. 32. A common nineteenth-century arrangement of a bed pushed into a corner surmounted by asymmetrical draperies. Green drapes over lace with red tassels and cords on the drapery and on the white lace bedspread. (Photo, Winterthur.)

FIG. 18. *Croisées à lambrequins brodés et à passementeries.* From Pinsonnière, ed., *Nouveau Recueil de draperies arrangé et lithographié par E. Julienne* (Paris: Pinsonnière, Successor of Osmont, 1839), pl. 38. Designs reminiscent of Louis XIV styles: *left,* puce over lace; *right,* chintz; both with passementerie-encrusted valances, gold ropes and tassels, and gilt tiebacks and pelmets. (Photo, Winterthur.)

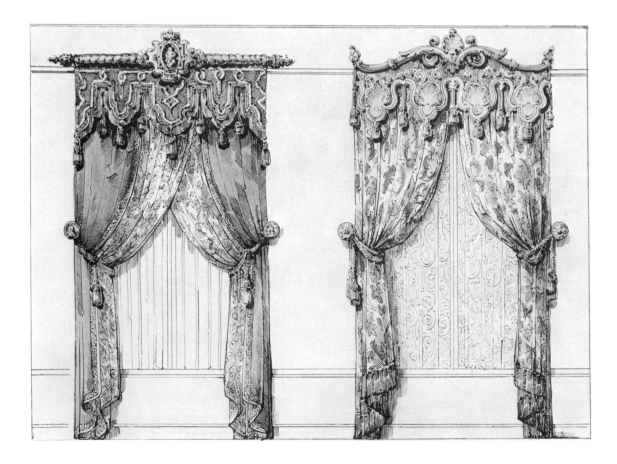

FIG. 19. *Ornements tirés de la collection des modèles de Mr. Pinsonnière*. From Pinsonnière, ed., *Nouveau Recueil de draperies arrangé et lithographié par E. Julienne* (Paris: Pinsonnière, Successor of Osmont, 1839), pl. 51. Plate showing the variety of drapery hardware available in 1839. (Photo, Winterthur.)

Fig. 20. Continuation of figure 19.

FIG. 21. Simplified window treatments and hardware for cottage dwellings. From J. C. Loudon, *An Encyclopaedia of Cottage, Farm, and Villa Architecture and Furniture* (London: Longman, Rees, Orme, 1833), pp. 338–39. (Photo, Winterthur.)

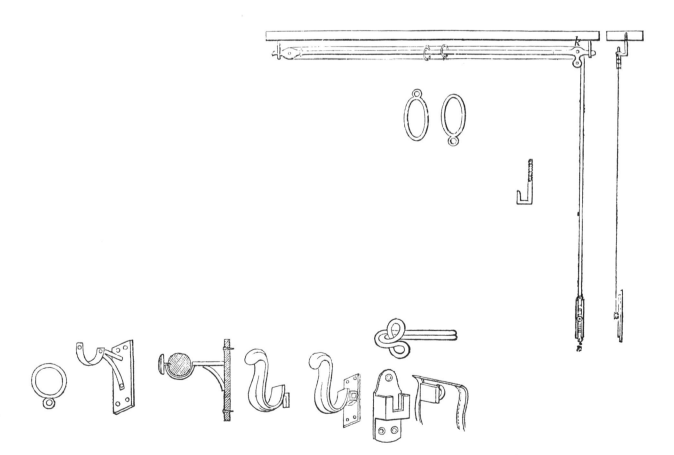

FIG. 22. *Décoration d'un fond de salon à trois croisées.* From *Le Garde-meuble, ancien et moderne* (Paris: D. Guilmard et Bordeaux frères, n.d.), pl. 142. Gold draperies trimmed with purple fringe, galloon, ropes, and tassels, and suspended from a gilt pole over lace curtains. Upholstery on the chairs and benches matches the drapery trim. (Photo, Winterthur.)

FIG. 23. *Décor de salle à manger avec croisée à trois battante.* From *Le Garde-meuble, ancien et moderne* (Paris: D. Guilmard et Bordeaux frères, n.d.), pl. 216. Red dining room drapery trimmed in green with banners of drapery material beside the window. (Photo, Winterthur.)

FIG. 24. *Décor de fenêtre*. From *Le Garde-meuble, ancien et moderne* (Paris: D. Guilmard et Bordeaux frères, n.d.), pl. 260. Two-tone green damask with trimmings to match complimented by a central panel of green velvet. This plate was the basis for drawing room draperies in the Gallier house, New Orleans, which is currently being restored. (Photo, Winterthur.)

FIG. 25. *Décor de fond de salle à manger.* From *Le Garde-meuble, ancien et moderne* (Paris: D. Guilmard et Bordeaux frères, n.d.), pl. 298. A triple window arrangement of gold, green, and terra-cotta Roman stripes, with gold and terra-cotta trimmings over lace curtains. The painted shades carry hunting and fishing motifs considered appropriate to a dining room. (Photo, Winterthur.)

FIG. 26. Parlor windows. From *Godey's Lady's Book* 49 (Aug. 1854): 97. Godey republished this same design in August 1858. (Photo, Winterthur.)

FIG. 27. Parlor draperies. From *Godey's Lady's Book* 49 (Aug. 1854): 170. (Photo, Winterthur.)

FIG. 28. Window curtain fashions. From *Godey's Lady's Book* 58 (Sept. 1859): 193. (Photo, Winterthur.)

FIG. 29. *Décor de cheminée*. From A[ntonio] Sanguinetti, *L'Ameublement au XIX siècle* (Paris: Au Bureau du Journal le Moniteur, [ca. 1863]), pl. 23. A design for a draped mantel and mirror first mentioned as "a recent innovation" in *Godey's Lady's Book* 42 (Oct. 1851): 243. (Photo, Vannucci Foto-Services.)

FIG. 30. *Croisée de salle à manger*. From A[ntonio] Sanguinetti, *L'Ameublement au XIX siècle* (Paris: Au Bureau du Journal le Moniteur, [ca. 1863]), pl. 72. A design for a lambrequin or *cantonnière*-type dining room drapery over the perennially popular painted shade. (Photo, Vannucci Foto-Services.)

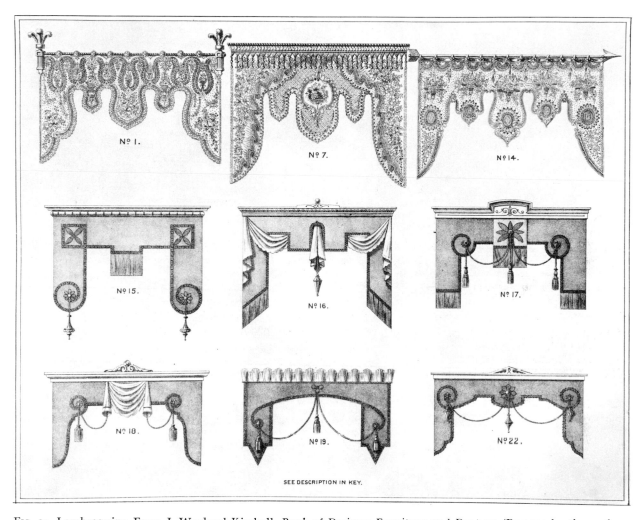

FIG. 31. Lambrequins. From J. Wayland Kimball, *Book of Designs: Furniture and Drapery* (Boston: by the author, 1876), pl. 1. Of the nine lambrequins, numbers 1, 7, and 14 are intended to be made in lace and lined with white or colored cambric to match nearby furnishings, while the remaining designs represent simpler, less costly treatments. (Photo, Winterthur.)

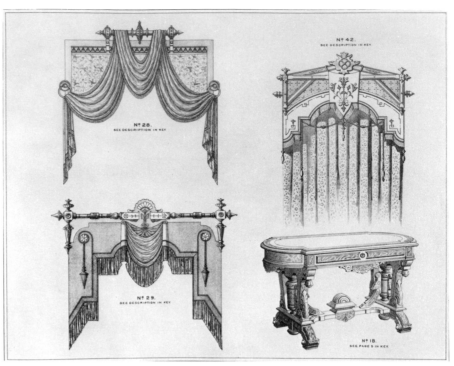

Fig. 32. Lambrequins. From J. Wayland Kimball, *Book of Designs: Furniture and Drapery* (Boston: by the author, 1876), pl. 20. *Top left,* cornice covered with plush, framing a piece of figured tapestry or silk as a background for a full drapery of plain material trimmed with gimp and fringe; *bottom left,* ebonized gilt wood cornice with drapery in plain material possibly using two colors; *top right,* black walnut cornice, carved and lined, with leather valance in two shades and figured plush curtain. (Photo, Winterthur.)

Fig. 33. Windows. From J. Wayland Kimball, *Book of Designs: Furniture and Drapery* (Boston: by the author, 1876), pl. 21. *Right,* lambrequin "should be made of a plain material and is especially designed for two colors. . . . the upperplain part may be of drab, the fulled piece festooned across . . . blue. The fulled under-valance or wings . . . drab, and the stiff pipes from which the hangars drop . . . blue. . . . The cords, fringes, and trimmings . . . blue, black, and gold." (Photo, Winterthur.)

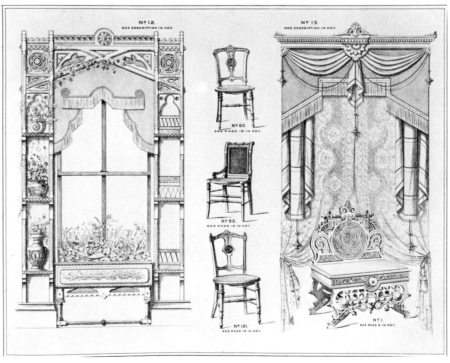

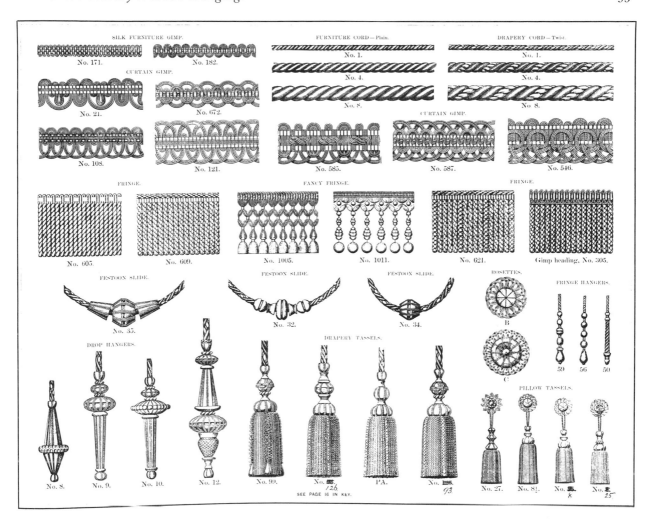

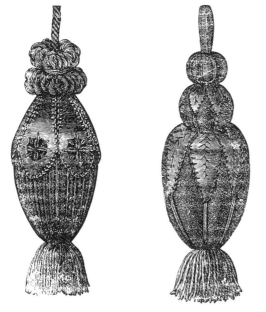

FIG. 34. Details of fashionable trimmings. From J. Wayland Kimball, *Book of Designs: Furniture and Drapery* (Boston: by the author, 1876), pl. 27. (Photo, Winterthur.)

FIG. 35. Silk tassels. From C. S. Jones and Henry T. Williams, *Beautiful Homes: How to Make Them* (Rockford, Ill.: Rockford Publishing Co., 1885), p. 64. Instructions for making the tassels are included. (Photo, Winterthur.)

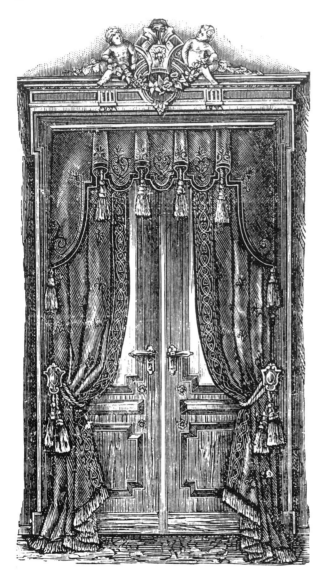

FIG. 36. Portiere and lambrequin. From C. S. Jones and Henry T. Williams, *Beautiful Homes: How to Make Them* (Rockford, Ill.: Rockford Publishing Co., 1885), p. 58. "Made of brown woolen reps lined with muslin; edges bordered in dark brown and black velvet applique and embroidery worked on a foundation of light brown cloth; application edged with yellow silk soutache; satin stitch design figures worked with brown sadler's silk in two shades and edged with yellow cord fastened on black silk; knotted stitches worked with gold thread." (Photo, Winterthur.)

FIG. 37. Portiere made from an oriental rug. From *Decorator and Furnisher* 3, no. 3 (Dec. 1883): 98. (Photo, Winterthur.)

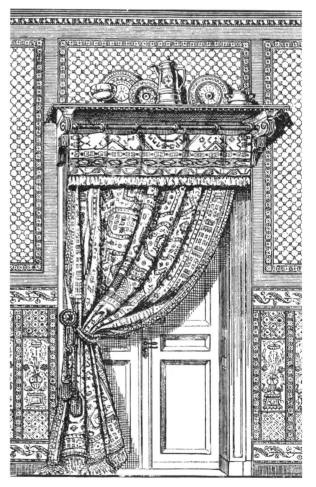

FIG. 38. Window arrangement. From Eugene Plasky, *La Tenture aristocratique* (n.p., 1880s), pl. 45. (Photo, Vannucci Foto-Services.)

FIG. 39. Installation chart. From Eugene Plasky, *La Tenture aristocratique* (n.p., 1880s), pl. 45. (Photo, Vannucci Foto-Services.)

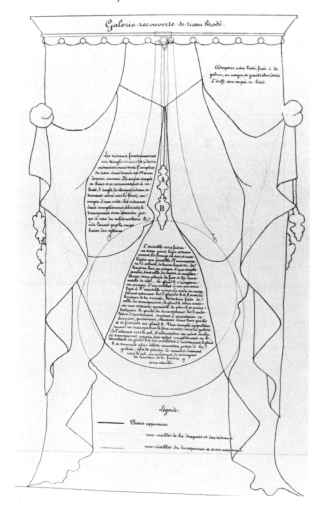

98

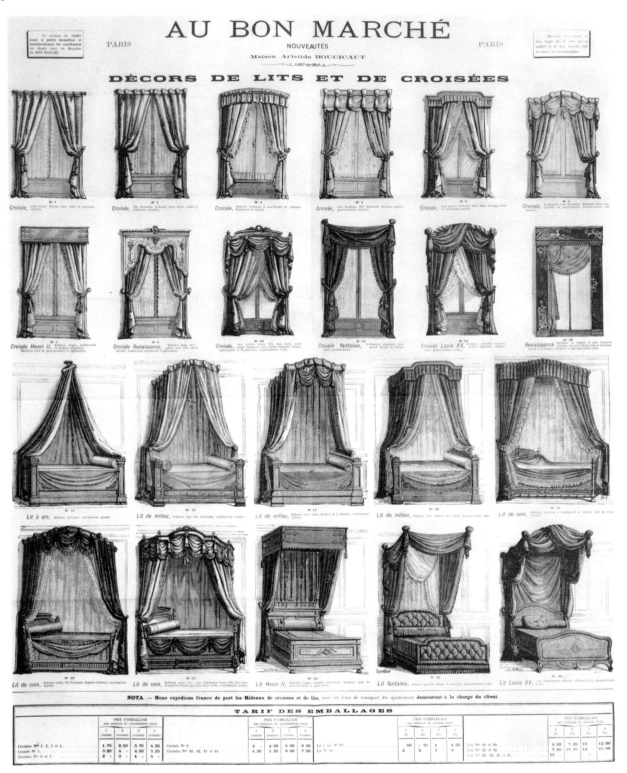

FIG. 40. Bed and window draperies. From advertisement for Au Bon Marché, Paris, Feb. 1893. Note the styles in use for over forty years. (Photo, Winterthur.)

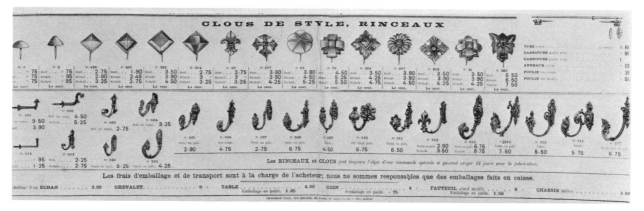

Fig. 41. Drapery ornaments. From advertisement for Au Bon Marché, Paris, Nov. 1892. Hooks, tiebacks, and ornamental nailheads. (Photo, Winterthur.)

# Associated Artists and the American Renaissance in the Decorative Arts

*Wilson H. Faude*

THE UNITED STATES, in the 1870s and 1880s, was enjoying the benefits of its great industrial and commercial expansion, and with these profits came the demand, especially from the newly rich, for luxuries of all kinds. Despite its newly found economic strength and the increasing output of its mills and factories, the country was still dependent on European trends and influences for inspiration in the arts of design. The fine decorative arts that enhanced the homes of the wealthy were primarily of foreign design, if not also of foreign manufacture. The expanding middle-class market welcomed factory-produced designs based on European styles, but the wealthy classes, who fancied themselves the tastemakers of the nation, sought the status attached to imported goods and lamented the lack of original American designs of quality. Decorating firms, too, succumbed to the aura of European infallibility. One firm, Cheney Brothers, in Manchester, Connecticut, sent buyers abroad to purchase samples of German, French, and English silks and prints, and employed a German draftsman to adapt them to American machines. The country's inability to produce original fine designs for the domestic market was noted at the Philadelphia Centennial Exhibition of 1876. Walter Smith, in his review of the exhibition, stated that foreign manufactures were "competing more successfully with native products . . . monopolizing the vast sums annually expended by the ever-increasing wealthy classes." [1]

In the five years from 1876 to 1881, America witnessed a renaissance in the decorative arts. The rebirth of fine design resulted in the rejection of many previously accepted foreign styles and in the development and popularity of designs and styles fashioned by American firms. The impetus for this change came from American manufacturers of expensive decorative-arts goods and from their well-to-do clients. The rapid popular acceptance of the new American designs may have resulted from the international awards they received, but, more basically, this success can be attributed to the creative artists who were challenged by home decoration as a serious and worthwhile discipline. Paramount in this group was the professional New York decorating firm of Associated Artists.

Associated Artists was created in 1879 by the partnership of Louis Comfort Tiffany (1848–1933), Samuel Colman (1832–1920), Lockwood de Forest (1850–1932), and Candace Thurber Wheeler (1828–1923). In the four years of their partnership, the firm captured many of the prestigious commissions of the day, including the redecoration of the White House in 1882. Each commission was approached with a freshness and an ingenuity that Cecelia Waern, an English visitor in the 1890s, likened to "a clever milliner adapting adroitly to any problem presented." Some critics objected to the firm's work; one writer for *Scribner's Monthly* categorized the firm's interiors as disregarding the traditions of all the great decorators who used harmonies of red, blue, and gold. In the view of that critic and others, Associated Artists was merely providing "an improvement upon the color-treatment of the Saracens." [2]

[1] Candace Wheeler, Document no. 1, n.d., Mark Twain Memorial Collection (hereafter MTMC), loan of Mrs. William T. Pullman, to whom the author is indebted for granting permission to use this collection of letters and for providing related information on Candace Wheeler; Smith, *Examples of Household Taste* (New York: R. Worthington, 1876), p. 498.

[2] Waern quoted in Robert Koch, *Louis C. Tiffany, Rebel in Glass* (New York: Crown, 1964), p. 13; William C. Brownell, "Decoration in the Seventh Regiment Armory," *Scribner's Monthly* 22, no. 3 (July 1881): 373.

Louis Comfort Tiffany, the son of Charles Lewis Tiffany, founder of the New York jewelry firm, Tiffany and Company, began his career as a painter, studying under George Inness and Samuel Colman. He traveled abroad with Colman, particularly in North Africa and Europe, and in 1877, with John La Farge and others, he founded the Society of American Artists. Although he is primarily remembered for his designs in iridescent and stained glass, Tiffany exhibited a number of his paintings at the Philadelphia exposition in 1876; he was an early exponent of the belief that a credible artist should not confine himself to a single medium. Samuel Colman studied under Asher B. Durand from 1860 to 1862 and, from 1871 to 1876, traveled in Europe and North Africa. A founder and the first president of the American Watercolor Society, he was well known not only for his paintings and watercolors but also for his designs in fabrics and in wallpapers and ceiling papers. Lockwood de Forest, like Tiffany and Colman, began his career as a painter, studying with Herman Corrode, in Rome, and with Frederick Church and James M. Hart. From 1875 until 1878 he traveled in Europe, Egypt, Syria, and India. In 1881 he established workshops in Ahmadabad, India, employing from forty to one hundred men and boys of the Mistri caste, for the revival of Indian designs in wood and perforated brass (Fig. 1).[3] De Forest is perhaps best known today for his use of these materials in the design schemes of fashionable homes.

Candace Thurber Wheeler was neither an artist nor a professional designer, but a housewife, until she visited the Centennial Exhibition of 1876. There she saw the English exhibition of art needlework from the Kensington School, a school established to aid economically deprived gentlewomen. Realizing that such a society could provide work for thousands of helpless women who "were ashamed to beg and untrained to work," she established in New York, in 1877, the Society of Decorative Art. Prominent artists, including Tiffany, de Forest, Colman, and La Farge, were enlisted to teach classes in design. Over thirty sister societies were soon established in the United States, including ones in Boston, Philadelphia, and Hartford. Her concern for the education of women led Wheeler to write many books on home decoration,

including *The Development of Embroidery in America* (1894), *How to Make Rugs* (1900), *Principles of Home Decoration* (1903), and her autobiography, *Yesterdays in a Busy Life* (1918). By 1879 Candace Wheeler was well known as a designer of fine embroideries and textiles; she had won several competitions, and art journals featured her designs as models to be copied. That year Tiffany resigned from the Society of Decorative Art, explaining: "It's all nonsense this work. . . . You can't educate people without educational machinery, and there is so much discussion about things of which there is really no question." Tiffany proposed that Wheeler join Colman, de Forest, and himself in creating a professional decorating firm, a firm unlike the Society of Decorative Art or similar organizations. Tiffany had in mind "a business[,] not a philanthropy or an amateur educational scheme. We are going after the money there is in art, but art is there all the same." Originally the firm was to be called the Louis C. Tiffany Company, but at Wheeler's suggestion the name Associated Artists was adopted instead. To contemporaries, the names Louis C. Tiffany Company and Associated Artists were interchangeable.[4] In 1883, the partnership dissolved, but Candace Wheeler retained the name Associated Artists for her decorating firm, whose primary concern was the design of textiles and wallpapers and ceiling papers.

In the 1870s and 1880s there developed in England and America a genuinely modern style in the decorative arts that was quite separate from the heavily patterned, lambrequined design schemes usually associated with these decades. The aesthetic movement, as this ubiquitous style was labeled, beginning with the work of a few designers and architects, ultimately embraced every art form from architecture to greeting card design. It fostered a period of concern about interior design, a period during which people talked about, wrote about, and spent vast sums in cultivating their tastes. In England the movement was led by E. W. Godwin (1835–86), James Abbott McNeill Whistler (1834–1903), and Christopher Dresser (1834–1904). The American counterpart to the aesthetic movement was manifested in the work of Louis C. Tiffany and Associated Artists. The partners saw themselves as arbiters of good taste and, consequently, retained a tight aesthetic control over their projects.

---

[3] De Forest, *Illustrations of Design* (Boston: Ginn & Co., 1912), pp. iii–v.

[4] Wheeler, *Yesterdays in a Busy Life* (New York: Harper & Bros., 1918), pp. 231, 233.

FIG. 1. Squares designed and carved in teakwood by men of the Mistri caste. From Lockwood de Forest, *Illustrations of Design* (Boston: Ginn & Co., 1912), pl. 24. (Collection of Henry Darbee: Photo, Mark Twain Memorial.)

Unlike William Morris and the eloquent reforming handicrafters, Godwin, Whistler, Dresser, Tiffany, and Associated Artists pioneered designs for industry. Since they regarded medieval methods as inadequate to satisfy the needs of or to express the character of their culture, they designed wallpapers, textiles, and furniture for industry and for wealthy clients. Many of their designs were consciously adapted from native American flowers and plants. They introduced American glass, woods, textiles, and metals to the public in accordance with the highest standards of decorative art.[5] They also borrowed freely from oriental, Turkish, and Indian sources. As Japanese enthusiasts, they developed a daring use of pale tints in sparsely furnished interiors. Regrettably, black and white photographs merely preserve the patterns of these design schemes without conveying the harmony or feeling of the rooms: pale walls covered with metallic patterns, peacock blue heightened with gold, wallpapers highlighted with small gold sunbursts. Photographs cannot reproduce the melody of color and tone nor the refreshing interplay of light and texture of the designs. The work of Associated Artists was a study in subtle contrasts.

Each partner in Associated Artists was in charge of a specific department. Wheeler's was textiles and embroideries, Tiffany's glass, de Forest's carvings and wood decoration, and Colman was director of color.[6] All had traveled widely, and from their travels and their separate backgrounds they brought a variety of design ideas to the firm. Although they often turned to Indian or oriental motifs for inspiration, the success of their designs was due largely to their belief that good designs must be appropriate to the client and the setting, and that the designer must "have personal gifts of grace and composition, and an education which is not only technical, but special and literary." The greatest factor that brought recognition and prestigious commissions to Associated Artists was an exacting but pragmatic notion of beauty to which the firm consistently and successfully adhered. "The first principle of beauty is appropriateness, and no room could be beautiful which failed to express the individuality of the occupant." In creating appropriateness, attention to detail, both detail of workmanship and detail of design, was basic to

the firm's approach. Candace Wheeler explained: "It is a great mistake to suppose that in small things the rules of art, the philosophy of art, may be neglected. Small things cease to be unimportant if largely treated." Thus, the firm endeavored to harmonize all the decorative details, however minor, in a room. *Artistic Houses* and other critics agreed that not an article of furniture or ornamentation incorporated in a room designed by the firm "was put there save under the . . . exigency of the decorative scheme." [7]

This attention to detail was reflected in the firm's consistent philosophy of color treatment. Again Candace Wheeler best expressed the partners' views, which were applied in numerous commissions. "In applying principles of color . . . the first and most important one is that of gradation. The strongest, and generally the purest, tones of color belong . . . at the base; and the floor of a room means the base upon which the scheme . . . is to be built. . . . If a single tint is to be used, the walls must take the next gradation, and the ceiling the last." [8] Thus, in overall conception, ornamentation, and color, rather than slavishly copy European historical styles, the firm strove to adapt designs to American uses and materials, and attempted to present a complete, harmonious decorative effect that merged appropriateness and individuality.

The success of any interior-decorating firm is usually measured by three factors: first, the number of important commissions the firm receives; second, the degree to which other firms imitate its work; and, third, the publicity and criticism accorded its work by contemporary periodicals and journals. This paper will evaluate the work of Associated Artists in terms of the firm's important commissions. From the beginning, Associated Artists received such commissions. This immediate success may have been due to the social connections of the partners, their individual artistic reputations, or simply the willingness of well-to-do clients to gamble on the firm, hoping for the prestige of featuring a new style. The work of Associated Artists from 1879 to 1883 can be viewed in terms of the

---

[5] Constance Cary Harrison, *Woman's Handiwork in Modern Homes* (New York: Charles Scribner's Sons, 1882), p. 5.

[6] Wheeler, Document no. 1, n.d., MTMC.

[7] Wheeler, "The Philosophy of Beauty Applied to House Interiors," in *Household Art,* ed. Candace Wheeler (New York: Harper & Bros., 1893), p. 32; Wheeler, "Decorative and Applied Art," in *Household Art,* p. 202; *Artistic Houses,* 2 vols. (1883; reprint ed., 2 vols. in 1, New York: Benjamin Blom, 1971), 1, pt. 1: 54.

[8] Wheeler, "Philosophy of Beauty," p. 8.

FIG. 2. Associated Artists, embroidered drop curtain in the Madison Square Theatre. New York City, 1879–80. From an 1883 program. (Collection of Robert Koch.)

three types of commissions the firm executed: public or institutional, private, and commercial.

The first public commission completed by Associated Artists was the drop curtain for the new Madison Square Theatre (1879–80). The design for the curtain was adopted from a silk picture by Mrs. Oliver Wendell Holmes, Jr., which represented a vista in the woods (Fig. 2). It provided "wonderful studies of color . . . a daring experiment in methods of appliqué." Because the curtain covered some ninety yards, and because it was the firm's first public commission, all four partners were directly involved in its creation. Tiffany adapted the design and improvised ingenious expedients as to method; de Forest was in charge of materials, Colman of color, and Wheeler of the actual execution. Great care was taken to assure that the velvety leaves, the shiny leaves, the silky seedpods, and the fibrous stalks retained their natural qualities.[9] The *Art Interchange*, a popular periodical of the day, ac-

[9] Wheeler, *The Development of Embroidery in America* (1894; reprint ed., New York: Harper & Bros., 1921), pp. 124–25; Wheeler, *Yesterdays*, p. 233.

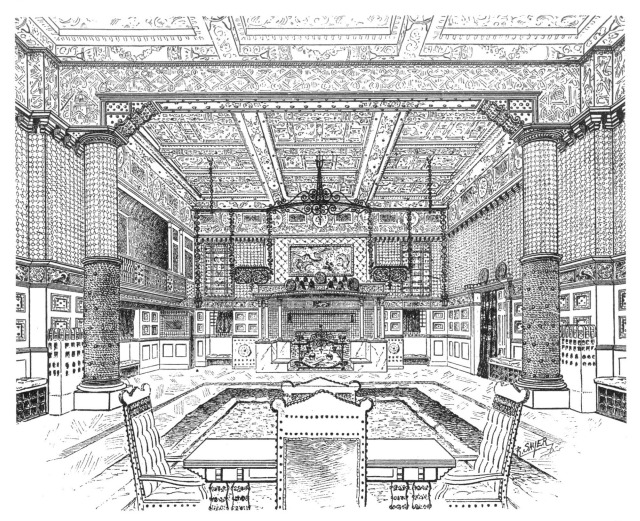

FIG. 3. Associated Artists, Veterans' Room in the Seventh Regiment Armory. New York City, 1879–80. From *Scribner's Monthly* 22, no. 3 (July 1881): 371. (Photo, Mark Twain Memorial.)

corded the curtain a long review, stating that "its influence upon the needlework of America will [never] be entirely lost." The journal's description is worth quoting because it provides one of the best contemporary depictions of the firm's rich yet subtle palette. The curtain represented

a line of tangled weedy shore, where meadow grasses tall, wild lilies, bold looking black eyed Susans, daisies, rushes and arrow heads grow, while above them stretch branches of tulip trees in flower, and clinging vines, and shadowy boughs lead the eye into a misty background . . . the effect is given of sunlight breaking through . . . butterflies and bees in plenty are a-wing . . . and dew seems to glitter in the morning sun. A width of lustrous, water-lined blue, gives depth and contrast to the flower masses. . . . This part of the hanging is bordered with richest velvets and plushes, in tints which shade from

the line of gold, which separates them from the blue, through olive to black; the richness of the stuff being heightened by the introduction of discs of heavy gold and silk embroidery. At the top of the curtain this treatment is repeated in paler tints of blue and old gold.[10]

The curtain was in place for the theater's first play, *Hazel Kieke,* presented on February 4, 1880, but in March of that year it was destroyed by fire. An "improved copy" replaced it, but it too was eventually destroyed. The work represented the first time that a major curtain for an American public theater had been constructed entirely of appliquéd textiles. Its novelty and popularity added to the firm's reputation. Yet Associated Artists tried "not to repeat the

[10] *Art Interchange* 4, no. 5 (Mar. 30, 1880): 38, 30.

triumphs . . . but to see how far the best which had been done was applicable to the present." [11]

At this time, when the Madison Square Theatre was under construction, the Seventh Regiment Armory was being built on Park Avenue. The Seventh Regiment was New York's finest, and its ranks were filled with the best of New York society. Associated Artists was commissioned to decorate the library and the Veterans' Room (Fig. 3) of the new armory for a fee of $20,000. William C. Brownell described the project in an article for *Scribner's Monthly*. It was, Brownell observed, "the first of [the firm's] performances on a large scale." [12] Tiffany conceived the general character of the room, playing heavily on military themes. The ceiling was paneled in wood, with ponderous beams at periodic intervals. The oak wainscot stood some ten feet high, and in every detail, from the silver stenciling on the ceiling to the wrought iron chandeliers to the heavy furniture, the scale was one of military triumph.

In designing the room, Tiffany employed Stanford White as an architectural consultant. Colman was responsible for the oriental and eastern arabesque designs covering the walls and ceiling, as well as for the delicate color harmonies. The frieze, which represented chronological progress in the art of war, with shields and allegories interspersed with geometric motifs, was painted by Francis Davis Millet and George Henry Yewell. At the north end of the room stood a large fireplace, with its red brick surrounded by a bed of deep blue Tiffany glass tiles. Above the fireplace opening was a representational plaque of "an eagle swooping down upon a dragon, or octopus . . . lashing into foam the 'ignominious ooze.'" Candace Wheeler's portieres were cited by Brownell, and some years later by an author in *Harper's Monthly,* for their appropriateness as well as their skillful design and use of materials. The portieres were made of Japanese brocade representing leopard skin, with velvet appliqué depicting "the days of Knighthood and romantic warfare." The intermediate spaces of the portieres were covered with fine rings, suggesting medieval chain mail. [13]

A memorial booklet published for the dedication of the room praised the decorations, citing the "poetic adaptations of the decorative material to the purpose in hand," and declaring the created effect "undeniably assimilable . . . with the huge hard, clanging ponderosities of war and tramping regiments and armories." Brownell did not agree, though he openly admitted that his decorative tastes leaned toward an architect's point of view rather than a decorator's. He felt that only classical motifs in reds, blues, and golds were acceptable for a client as established as New York's Seventh Regiment, and he wrote that although the firm of Associated Artists might find contemporary decorations "barbaric and brutal," traditional color schemes avoided exactly what Tiffany and his partners had created. But Brownell's argument faltered when he took note of the firm's one goal in decorating the room, a goal in keeping with Associated Artists' belief that decorations should express the individuality of the client. His most telling observation was "You cannot help seeing that you are in a Veteran's Room." [14]

The third important public commission undertaken by Associated Artists was the Union League Club of New York. In October 1880, the club contracted with Associated Artists to decorate the halls and grand staircase of its new clubhouse at Fifth Avenue and Thirty-ninth Street, and to design and execute the draperies for "the main dining room, the halls, the picture gallery, meeting hall, and the dining rooms on the third and fourth floors." The contract for the draperies stipulated that all of the designs be original and exclusive to the club, and called for a payment of $3,710, with the finished goods to be delivered by December 15, 1880. [15]

In addition to Associated Artists, the Union League Club also contracted with John La Farge, Frank Hill Smith, Augustus Saint-Gaudens, and Will H. Low to undertake decoration of other areas in the new building. The resulting dissonance and lack of unity was described in the *Century Illustrated* as a "variety obtained by giving the ceiling to Mr. Whistler . . . and the walls to the decorator

[11] Wheeler, *Yesterdays*, p. 234; Wheeler, *Development of Embroidery*, p. 125.

[12] Brownell, "Seventh Regiment Armory," p. 370.

[13] Brownell, "Seventh Regiment Armory," pp. 370, 375–76; Mrs. Burton Harrison, "Some Work of the 'Associated Artists,'" *Harper's Monthly* 69, no. 411 (Aug. 1884): 350.

[14] Joseph Purtell, *The Tiffany Touch* (New York: Random House, 1971), p. 117; Brownell, "Seventh Regiment Armory," pp. 370, 375.

[15] "Agreement Regarding Draperies at the New Union League Club Building," Oct. 15, 1880, Union League Club Archives, New York. Guy St. Clair, librarian, and Edwin H. Pullman of the Union League Club were very helpful in providing information for this article.

Fig. 4. Candace Wheeler, fish curtain in the dining room of the Union League Club. New York City, 1880. From *Century Illustrated* 23, no. 5 (Mar. 1882): 748. (Photo, Mark Twain Memorial.)

Garibaldi." [16] The most surprising part of the entire affair was that Associated Artists agreed to an arrangement allotting it only a portion of a room or a wall. Perhaps the young firm did so because of the importance of the commission, the payment received, the expected publicity, or the fact that many of the club's members were friends of the partners. In any event, it was distinctly out of character and was a practice the firm never repeated.

Despite criticism of the lack of harmony in the decorations of the new clubhouse, two creations by Associated Artists received considerable comment. The first was the draperies designed and executed by Candace Wheeler for the main dining room (Fig. 4). The design was a sea motif, with a net containing fish. On one side a sea gull swooped down for its prey, while the other panel featured a large fish striving to free itself from the net. The cur-

tains were bordered with wide pieces of blue plush, suggesting the deep blue sea. The elegant yet comfortable effect of the curtain was described by a writer for *Harper's Monthly* as giving "solace for those pathetic wanderers from home compelled to seek the shelter of a club house." [17]

For the members, the most successful feature of the new clubhouse was Tiffany's stained glass window on the first floor landing of the main staircase (Fig. 5). The *Century Illustrated*'s critique included a sketch of the stairwell, but it ignored Tiffany's window and instead cited La Farge's windows as perhaps the finest to be seen in any interior. Yet it was Tiffany's window that gave the main floor a "touch not only of splendor, but of warmth and welcome," radiating color onto the stair and the lobby. One member described the atmosphere created by the window as so strong that many would "walk down from the higher floors . . . just to stop at the landing and look." The placement of the window was no accident but a result of the careful control of light and design. Due to the painterly training of Tiffany, de Forest, and Colman, the firm saw to it that "the tint of any particular room [was] chosen with reference not only to personal liking, but first of all to the quantity and quality of light which pervades it." [18] On February 2, 1931, the Union League Club moved again, and the old clubhouse on Fifth Avenue was demolished. With its demolition, most of the records and decorations were lost.

Following the assassination of President James A. Garfield, in 1881, President Chester A. Arthur refused to move into the White House until it had been renovated and redecorated. The privilege of redecorating the executive mansion was awarded to Associated Artists. This, the last public commission undertaken by the partnership, marked a peak in the work of the firm. Few commissions as important as the White House existed; any changes to its interiors would be noted and imitated by other decorators and clients. Whereas Associated Artists had sheathed the Veterans' Room of the Seventh Regiment Armory in iron and oak, for the White House the firm chose lighter, classical motifs, ap-

---

[16] "Some of the Union League Decorations," *Century Illustrated* 23, no. 5 (Mar. 1882): 746.

[17] Harrison, "Some Work of the 'Associated Artists,'" p. 350.

[18] *Century Illustrated* 23, no. 5 (Mar. 1882): 746; Will Irwin, Earl Chapin May, and Joseph Hotchkiss, *A History of the Union League Club of New York City* (New York: Dodd, Mead & Co., 1952), pp. 108–9); Wheeler, "Philosophy of Beauty," p. 9.

FIG. 5. Louis C. Tiffany, stained glass window in staircase of the Union League Club. New York City, 1880. From Will Irwin, Earl Chapin May, and Joseph Hotchkiss, *A History of the Union League Club of New York City* (New York: Dodd, Mead & Co., 1952), opp. p. 108. (Photo, Mark Twain Memorial.)

propriate to the formal dignity of the president's mansion.

Since the principal rooms of the White House were used at night, Associated Artists created designs and used materials adapted to illumination by gaslight. The East Room (Fig. 6), measuring some 80 by 40 feet, served as the reception room for great occasions; here the firm installed a sienna-colored Axminster carpet, which harmonized with the formal white and gold wall treatment. The ceiling was covered in silver leaf stenciled in a fine mosaic pattern, with the leaf reflecting the gold of the carpet and creating an overall effect of warm elegance appropriate to the room's function. The Red Room (Fig. 7) received a new mantel, and, in keeping with the firm's principle of color gradation, all the woodwork was toned from dark red at the bottom to a lighter shade at the top, bringing

FIG. 6. Associated Artists, East Room of the White House. Washington, D.C., 1882–83. From *Artistic Houses,* 2 vols. (1883; reprint ed., 2 vols. in 1, New York: Benjamin Blom, 1971), 2, pt. 1: opp. 100. (Photo, Mark Twain Memorial.)

it into harmony with the overall color scheme. The fireplace opening was surrounded by panels of Japanese leather and by glass tiles in tints of amber and red. Below the old mirror, the firm added a mosaic of glass studded with glass gems, which would reflect the light at night.

In the Blue Room (Fig. 8) Associated Artists installed a soft, blue gray carpet and formal gilt and satin upholstered furniture. There was a gradation of color from the blue gray of the carpet to silvers and blue whites in the ceiling. The ceiling featured "ovals of a silvery tone," with their centers decorated in colored metals. The frieze was a light silver and gray, the walls were tones of robin's egg blue, and fine silver lines enhanced the woodwork. Colored glass was incorporated into the existing mirrors, and great care was taken to assure that the new mantel lambrequin of blue and gold silk harmonized with the tone and texture of the room's curtains and furniture coverings. *Artistic Houses*

summed up the impact of the room as a "scintillating effect of great variety and brilliancy."

The most noted addition to the White House was Tiffany's "glass mosaic" partition separating the entrance foyer from the main corridor. Aside from its artistic merits, the 338-square-foot partition, by screening off the entrance foyer from the main corridor, provided privacy for the president and his family, who could now cross the main floor without being bothered by office seekers or the general public. The screen, composed of colored glass, featured traditional patriotic symbols. The ceiling of the corridor was decorated in gold and ivory, with rosettes made of Indian brass, and the walls were painted an olive gold to harmonize with the gold tones of the screen.[19]

During this period Associated Artists also undertook a great many private commissions. In review-

[19] *Artistic Houses* 2, pt. 1: 97–99.

FIG. 7. Associated Artists, Red Room of the White House. Washington, D.C., 1882–83. (Collection of the Library of Congress.)

ing the work of any decorating firm, it is important to consider the relationship between the client and the decorator: the extent of control enjoyed by the decorator and the extent of control or influence exercised by the client. Regrettably, research on Associated Artists has produced a specific statement of this relationship in only one case. Tiffany, in agreeing to decorate the home of Samuel L. Clemens (Mark Twain) in Hartford, outlined the job to be performed: "Walls—painted or papered *at our option,* ceiling painted." [20] It would have best suited the firm always to decorate as it saw fit, with the client required only to pay the bills.

---

[20] Tiffany to Clemens, Oct. 24, 1881, Mark Twain Papers, University of California at Berkeley (hereafter MTP), italics added. All material under copyright published by permission of Mr. Thomas Chamberlain and the Mark Twain Company.

The one instance of direct client involvement in decorating was outlined in a letter from Wheeler to Tiffany concerning the firm's work in 1882 on the Fifth Avenue house of Cornelius Vanderbilt II. The drawing room of the house was to have its walls covered in green plush, with a series of tapestries designed by Candace Wheeler's daughter, Dora Wheeler, as the major decoration. The mythological subjects of the hangings featured *The Winged Moon* (Fig. 9), *The Water Spirit, The Birth of Psyche, The Air Spirit,* and *The Flower Girl.* They were done on "salmon pink stuff" in tones of blue, gray, and purple, with particular attention to subtle atmospheric effects. When the room neared completion, with curtains, window draperies, and wide portieres in place, Mrs. Vanderbilt "absolutely refused to have the green plush." Colman was called in, and he decided to

FIG. 8. Associated Artists, Blue Room of the White House. Washington, D.C., 1882–83. (Collection of the Library of Congress.)

FIG. 9. Dora Wheeler, *The Winged Moon*. New York City, 1882. Executed in needle-woven tapestry by Associated Artists. From Candace Wheeler, *The Development of Embroidery in America* (1894; reprint ed., New York: Harper & Bros., 1921), opp. p. 122. (Photo, Mark Twain Memorial.)

use "the pale tint which was on the walls." [21] Due to the lack of further documentation, Colman's response to the problem cannot be evaluated. The use of the pale tint instead of green plush might have been either an unfortunate compromise or a creative solution to meet the client's objection.

In its private work, the firm's commitment to relating design to a room's function in terms of the client's individuality was one of its strongest assets. Wheeler explained the firm's philosophy as follows: "Each room is in a certain sense the home of the individual occupant, almost the shell of his or her mind. There will be something despotic in the

house rules if this is not expressed." [22] The orientation of the rooms, their natural and artificial lighting, and their function determined the nature of each decorating scheme. The philosophy and general principles observed by Associated Artists in its private commissions are reflected in the partners' statements about selected rooms and halls, and in descriptions of various decorating tastes.

The drawing room or parlor of a Victorian house was usually formal in setting and atmosphere. Associated Artists felt that it was important to retain

[21] *Art Amateur* 10, no. 2 (Jan. 1884): 42; Wheeler to Tiffany, Jan. 28, [1883], courtesy of Mitchell Family Papers, Yale University Library.

[22] Wheeler quoted in Mary Gay Humphreys, "The Parlour," *Decorator and Furnisher* 2 (May 2, 1883): 53. Source and information extended to the author by Catherine L. Frangiamore, assistant curator, Cooper-Hewitt Museum of the Smithsonian Institution, New York.

FIG. 10. Associated Artists, drawing room of the George Kemp house. New York City, 1879. From *Artistic Houses*, 2 vols. (1883; reprint ed., 2 vols. in 1, New York: Benjamin Blom, 1971), 1, pt. 1: opp. 54. (Photo Robert Koch.)

this atmosphere of "delicacy and elegance" and recommended that "the walls . . . be hung with paper . . . or a scattered design in gold or silver." The color might be "blue or green or rose or cream," but the strongest consideration was that the end result be "delicate." [23]

Associated Artists' sense of color harmony, its eclectic adaptation of foreign designs to American uses, and its commitment to expressing both room function and individuality are illustrated in an 1879 commission to decorate George Kemp's apartment in New York. The decorations for the Kemp drawing room (Fig. 10) were described in *Artistic Houses* as a "delicious melody of color." The general motif of the room derived from Arabic designs, with the ceiling decorations in iridescent tones on a silver ground. The cornice receded to lighten the

scheme, and all of the woodwork, the casings, the chairs, the tables, and even the grand piano were specially designed to harmonize with the wall decorations and fabrics. Lace curtains, which had long been criticized for casting too strong a light, were replaced by curtains of Indian muslin or of sheer tambour with fine embroidered designs. The contemporary success of the design is indicated by the fact that in her book, *Woman's Handiwork in Modern Homes,* Constance Cary Harrison used the Kemp drawing room, with full color illustrations of the treatment of draperies and portieres, as the model to follow in creating interiors of good taste.[24]

The drawing room of the Samuel L. Clemens home in Hartford (Fig. 11), originally decorated in

---

[23] Wheeler, "Philosophy of Beauty," p. 32.

[24] *Artistic Houses* 1, pt. 1: 54; Harrison, *Woman's Handiwork,* pp. 186–89.

Fig. 11. Associated Artists, drawing room of the Samuel L. Clemens house. Hartford, Conn., 1881 (modern restoration). (Photo, Mark Twain Memorial.)

1874 in mid-Victorian style, with what one contemporary described as an "awful Victorian, rosebud carpet and blue satin curtains [and] . . . all that went with it," was redecorated in 1881 by Associated Artists. The walls were painted a pale pink and the woodwork ecru; both were stenciled in silver. Lily Gillette Foote, a close friend of the Clemens children, described the redecoration as transforming the room from "dark decorations and furnishings to something very much lighter and in the French style." The redecorated drawing room was described by Clara Clemens as presenting "an impression of hospitable *light*—of a suggestively divine quality." [25]

In decorating Hamilton Fish's drawing room (Fig. 12), the firm used a color scheme of blue and old ivory. The wainscot was peacock plush, paneled in wood, and the frieze was a low relief of bronze and delicate colors in a Moorish design. The existing mantelpiece was incorporated into the scheme by enclosing it in teak panels of East Indian design. The portieres were of blue silk, with the lower portion carrying the design of the wainscoting, and the upper panels representing delicate embroidered flowers. *Artistic Houses* commented that the room presented "an ascending scale of color from floor to ceiling." [26] For John Taylor Johnston's drawing room (Fig. 13), the color scheme chosen was salmon red, yellow, and brown. The fireplace opening was surrounded by Tiffany glass tiles and East Indian panels; the mirror had a border of colored glass repeating the Indian designs. The ceiling was painted, although the paint was applied with a palette knife "to give the effect of low relief." [27]

Unlike the varied social functions of the drawing room or parlor, the function of the dining room centered on food and conversation. Associated Artists felt that it was important to use a "pervading color" in the dining room so that "the walls and ceiling [would] be kept together by the use of one color only." [28] In decorating George Kemp's dining room (Fig. 14), the firm retained the original oak paneling and incorporated the wood's rich deep

tone into the color scheme. Above the paneling, Tiffany painted a harvest still life on gilded canvas. For the doors, he designed stained glass transoms that reduced and controlled the natural light. The draperies and hangings of embroidered plush were described by *Artistic Houses* as "the most beautiful specimens of pure and picturesque embroidery ever wrought in this country." [29]

Tiffany painted a similar frieze of fruits and vegetables over the mantel in the dining room of his own apartment (Fig. 15). There he employed embroidery on cloth in the frieze and in the lower band on the wall, with Japanese mushroom paper on the ceiling completing the decorative scheme. Japanese papers were also used to cover the walls and ceiling of Dr. William T. Lusk's dining room. To control the light in that room, the firm designed large transoms of stained glass in amber tones. This reduced the full-length windows some 3 feet, creating an intimate effect.

For Samuel Clemens's dining room, the clear finished woodwork was stenciled in gold, its pattern suggestive of Japanese designs (Fig. 16). The embossed wallpaper, with a design of lilies rampant, was finished in lacquered red and gold (Fig. 17). The dining room, with its northern exposure, had a pervading color scheme of red, gold, and brown, in exact agreement with the firm's philosophy that "if it has a cold northern exposure, reds or gold browns are indicated." [30] The north window of the dining room was directly over the fireplace. Tiffany preserved this feature but redesigned the mantel and its surround to bring it into harmony with the new decorations and to control the natural light. Tiffany used his opaque blue and amber glass tiles to form the fireplace surround and placed a brass hood inset with translucent amber glass tiles in the chimney opening. In the window above the mantel, surrounding a large light of clear beveled plate glass, were set translucent panels of amber glass framed by a panel of blue glass. On either side of the window, set in the wall, were two more stained glass panels, with subtle gradations of tint from gold below to blue above.

Because a library is less formal than a drawing room, the firm advocated that its color scheme be "much warmer and stronger than that of a parlour." If the room had a southern exposure, the

[25] Harriet Foote Taylor to Edith Colgate Salsbury, May 3, 1959, MTMC; Lilly Gillette Foote, "Reminiscences," 1958, MTMC; Clara Clemens to Robert H. Schutz, Feb. 21, 1959, MTMC.

[26] *Artistic Houses* 2, pt. 1: 95.

[27] *Artistic Houses* 2, pt. 2: 153.

[28] Wheeler, "Philosophy of Beauty," pp. 23–24.

[29] *Artistic Houses* 1, pt. 1: 53–56, 4.

[30] Wheeler, "Philosophy of Beauty," p. 24.

FIG. 12. Associated Artists, drawing room of the Hamilton Fish house. New York City, ca. 1883. From *Artistic Houses,* 2 vols. (1883; reprint ed., 2 vols. in 1, New York: Benjamin Blom, 1971), 2, pt. 1: opp. 95. (Photo, Mark Twain Memorial.)

FIG. 13. Associated Artists, drawing room of the John Taylor Johnston house. New York City, ca. 1883. From *Artistic Houses,* 2 vols. (1883; reprint ed., 2 vols. in 1, New York: Benjamin Blom, 1971), 2, pt. 2: opp. 151. (Photo, courtesy New-York Historical Society.)

FIG. 14. Associated Artists, dining room of the George Kemp house. New York City, 1879. From *Artistic Houses,* 2 vols. (1883; reprint ed., 2 vols. in 1, New York: Benjamin Blom, 1971), 1, pt. 1: opp. 54. (Photo, courtesy New-York Historical Society.)

FIG. 15. Associated Artists, dining room of Louis C. Tiffany's apartment near Madison Square. New York City, 1879. From *Artistic Houses,* 2 vols. (1883; reprint ed., 2 vols. in 1, New York: Benjamin Blom, 1971), 1, pt. 1: opp. 4. (Photo, Robert Koch.)

Fig. 16. Associated Artists, detail of the gold stenciled woodwork in the dining room of the Samuel L. Clemens house. Hartford, Conn., 1881 (modern restoration). (Photo, Mark Twain Memorial.)

Fig. 17. Associated Artists, detail of the embossed paper, finished in lacquered red and gold, in the dining room of the Samuel L. Clemens house. Hartford, Conn., 1881 (modern restoration). (Photo, Mark Twain Memorial.)

Fig. 18. Associated Artists, library of the Samuel Colman house. Newport, R.I., ca. 1883. From *Artistic Houses*, 2 vols. (1883; reprint ed., 2 vols. in 1, New York: Benjamin Blom, 1971), 2, pt. 1: opp. 71. (Photo, Robert Koch.)

firm recommended "green or strong India blue." [31] Samuel Colman decorated his own Newport library (Fig. 18) in a tone of blue black; the ebony woodwork, fine Japanese silks, and the ceiling with a Moorish design created a "stunning effect." Colman owned a vast collection of fine fabrics, many of which he incorporated into the firm's design schemes. To one critic, Colman's library, with its subtle play of design and texture, was a "perpetual feast." [32]

In the library of the Kemp house (Fig. 19), the firm covered the walls with silk, using embroidered plush for the frieze. The velvet draperies were highlighted with galloons, and again stained glass transoms were introduced. Against the gilt cove of the lightly painted and paneled ceiling, Associated Artists placed iridescent seashells, which lightened the ceiling and gave it sparkle and the desired richness. In W. S. Kimball's library (Fig. 20), in Rochester, New York, Tiffany glass tiles were again used in the fireplace surround. The heavy mahogany mantel and stone arch were lightened by the stained glass windows flanking the fireplace within the arch. The firm preserved the mahogany woodwork of the bookcases, trim, and ceiling, yet brought the decorations into harmony by painting the walls and ceiling and stenciling them with a fine Moorish design.[33] Samuel Clemens's library

[31] Wheeler, "Philosophy of Beauty," p. 24.
[32] *Artistic Houses* 2, pt. 1: 73.

[33] *Artistic Houses* 2, pt. 1: 73, pt. 2: 160.

FIG. 19. Associated Artists, library of the George Kemp house. New York City, 1879. From *Artistic Houses,* 2 vols. (1883; reprint ed., 2 vols. in 1, New York: Benjamin Blom, 1971), 1, pt. 1: opp. 55. Note the ceiling of iridescent seashells. (Photo, courtesy New-York Historical Society.)

FIG. 20. Associated Artists, library of the W. S. Kimball house. Rochester, N.Y., ca. 1883. From *Artistic Houses,* 2 vols. (1883; reprint ed., 2 vols. in 1, New York: Benjamin Blom, 1971), 2, pt. 2: opp. 160. (Photo, courtesy New-York Historical Society.)

FIG. 21. Associated Artists, mantel in the library of the Samuel L. Clemens house. Hartford, Conn., ca. 1880. From *Harper's Monthly* 71, no. 425 (Oct. 1885): 725. (Photo, Mark Twain Memorial.)

was dominated by a great Scottish mantel (Fig. 21), clear-finished mahogany bookcases, and, at the south end, a large conservatory (Fig. 22). In decorating the room, Tiffany proposed "walls covered with metal leaf stenciled, ceiling covered with metal leaf and paint." [34] The resulting walls were decorated in peacock blue stenciled in gold, the pattern suggestive of a Scotch plaid. The ceiling was also stenciled in gold.

The treatment of the front hall, the first area of a house seen by a visitor, was crucial. Halls of Victorian houses often appeared dark, due to heavy, dark-colored woodwork and the absence of windows. Associated Artists recommended red as the

most appropriate color for a hall, particularly Pompeian or Damascus reds, not only for their richness but also because these colors had enough yellow in their composition to harmonize with the yellows of oiled wood.[35]

For the front hall of the Clemens house (Fig. 23), Tiffany proposed: "walls painted and stenciled, ceiling painted and stenciled in metals, woodwork decorated or not, at our option; halls above the first floor, walls and ceiling plainly painted." The existing clear-finished walnut woodwork was incorporated into the scheme by stenciling a pattern in silver on it, which not only lightened the woodwork but gave it elegance, "as if inlaid with mother-of-pearl." [36] The walls and ceiling were painted red and stenciled with designs in black and silver. Teak carvings of Indian design were added to the walnut fireplace. In 1883, after the firm had completed the decorations, Samuel Clemens decided to alter the hall fireplace slightly by painting the white marble surrounding the opening "the same *strong red* of the hall walls, & then [covering] it with Mr. de Forest's thin arabesque—cut brass sheets, which will let the red show through" (Fig. 24).[37] At the landing between the first and second floors (Fig. 25), the color scheme changed from red with black and silver designs below to olive green and gold above. The effect of the hall treatment was described as having a "sense of brightness even on [a] dark day." [38] The hall of the W. S. Kimball house was described in *Artistic Houses* as appearing twice its actual size as a result of the firm's treatment of perspective and color. A large screen of East Indian teak panels shut off the staircase and organ loft from the hall. The openness of the carved panels created "an impression of great distance and mystery." [39]

Each interior by Associated Artists was unique, yet all were consistent with the firm's philosophy of beauty, color, harmony, function, and appropriateness to the owner. Many standard decorating features of the period were replaced, supplanted, or discarded by the firm. Lace curtains were replaced

[34] Tiffany to Clemens, Oct. 24, 1881, MTP.

[35] Wheeler, "Philosophy of Beauty," p. 21.

[36] Tiffany to Clemens, Oct. 24, 1881, MTP; Will M. Clemens, *Mark Twain, His Life and Works* (New York: F. Tennyson Nelly, 1894), p. 167.

[37] Samuel Charles Webster, ed., *Mark Twain Business Man* (Boston: Little, Brown & Co., 1946), p. 209, italics original.

[38] Clara Clemens to Schutz, Feb. 21, 1959, MTMC.

[39] *Artistic Houses* 2, pt. 2: 159.

Fig. 22. Associated Artists, library of the Samuel L. Clemens house. Hartford, Conn., 1881 (modern restoration). (Photo, Mark Twain Memorial.)

by Indian muslin, and traditional color schemes were abandoned. Stained glass controlled the light, and seashells became essential features of a design. In all, the decorator was in complete control. The lasting satisfaction that clients found in the work of Associated Artists was perhaps best expressed by Samuel Clemens in 1895, some fourteen years after the work had been completed. "How ugly, tasteless, repulsive, are all the domestic interiors I have ever seen in Europe compared with the perfect taste of this ground floor, with its delicious dream of harmonious color, and its all-pervading spirit of peace and serenity and deep contentments." [40]

The success of the firm and the desire of Amer-

[40] Dixon Wecter, ed., *The Love Letters of Mark Twain* (New York: Harper & Bros., 1949), p. 312.

ican manufacturers for American designs resulted in several commercial commissions. In 1879, Tiffany and Colman agreed to design wallpapers and ceiling papers for Warren, Fuller and Company. The company hired the art critic Clarence Cook to write the copy for their promotional booklet, *What Shall We Do with Our Walls?*, which was illustrated with the Tiffany and Colman wallpapers. In the pamphlet Cook stressed the need for well-designed wallpapers and ceiling papers. He attributed the lack of good designs to the "want of decorative talent among our house-painters." Ceiling papers were needed because "white ceilings are, or ought to be, disagreeable to everybody," and the work of professional fresco painters was "always irredeemably bad." The *Art Amateur* commented that "for the first time in this country, American

FIG. 23. Associated Artists, front hall of the Samuel L. Clemens house. Hartford, Conn., 1881 (modern restoration). (Photo, Mark Twain Memorial.)

FIG. 24. Detail of the front hall fireplace in the Samuel L. Clemens house, showing thin, cut brass sheets by Lockwood de Forest over the red painted marble. Hartford, Conn., 1881 (modern restoration). (Photo, Mark Twain Memorial.)

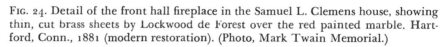

FIG. 25. Associated Artists, landing between the first and second floors of the Samuel L. Clemens house. Hartford, Conn., 1881 (modern restoration). (Photo, Mark Twain Memorial.)

artists of established reputation were devoting their talents to the designing of wallpapers." [41]

Tiffany's designs included one of wild clematis and cobwebs (Fig. 26). The clematis was rendered in gold, the cobwebs in silver, on a creamy yellow ground. Interspersed throughout the pattern were insects in metallic tints and, for variety, dashes of metallic colors. Another Tiffany design was a ceiling paper consisting of realistically rendered snowflakes colored to give the impression of abnormal height, suggesting the Milky Way. Colman's designs were more traditional in their division of the paper, indicating wainscoting, field, and frieze.

The maple leaf pattern (Fig. 27) was printed in gold on a plum-colored ground. The frieze showed Colman's use of graceful curves, suggestive of conventionalized Japanese clouds. Where the overall tone of the ground was plum colored, it was in ascending tones. The wainscoting was highlighted with crystals of fish scales.[42]

For a commercial firm to hire a renowned art critic and prominent artists to create an original style was not new, yet the zeal with which the firm of Warren, Fuller and Company fostered "a breach in the wall of old ideas and fashions of the past . . . to create something [with] an unborrowed, individ-

[41] *Art Amateur* 13, no. 1 (June 1880): 12; Clarence Cook, *What Shall We Do with Our Walls?* (New York: Warren, Fuller & Co., 1881), p. 30.

[42] Mary Gay Humphreys, "The Progress of American Decorative Art," in Wheeler, *Household Art,* pp. 153–54; *Art Amateur* 13, no. 1 (June 1880): 12.

FIG. 26. Louis C. Tiffany, wallpaper design. From Clarence Cook, *What Shall We Do with Our Walls?* (New York: Warren, Fuller & Co., 1881), opp. p. 18. (Photo, Mark Twain Memorial.)

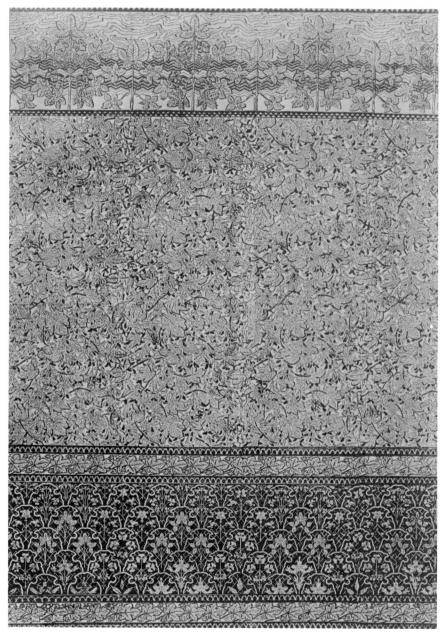

Fɪɢ. 27. Samuel Colman, *Parlor-Decoration* (wallpaper design). From Clarence
Cook, *What Shall We Do with Our Walls?* (New York: Warren, Fuller & Co., 1881),
opp. p. 24. (Photo, Mark Twain Memorial.)

ual look" was exceptional.[43] Throughout the 1880s
the firm sold papers designed "by American artists," headlining the work of Tiffany and Colman
and the Indian designs of de Forest (Fig. 28). These
designs were primarily aimed at the upper middle
classes and, although available commercially, had
no real impact upon the middle and lower classes.

[43] Cook, *What Shall We Do with Our Walls?*, p. 33.

In 1881, Warren, Fuller and Company sponsored
a competition for wallpaper and ceiling paper designs. Competitions in America were popular; design competitions for Christmas cards, furniture,
and even stoves were so numerous that one critic
commented, "If this country does not produce good
decorative artists it will certainly not be from the
lack of pecuniary incentive." Warren, Fuller and
Company offered $2,000 as prize money. The
judges were Christian Herter, Edward C. Moore,

FIG. 28. Warren, Fuller and Lange, advertisement featuring the designs of Louis C. Tiffany, Samuel Colman, and Lockwood de Forest. From *Art Amateur* 12, no. 4 (Mar. 1885): back cover. (Photo, Mark Twain Memorial.)

and Francis Lathrop.[44] Candace Wheeler and three of her associates in the embroidery studio of Associated Artists submitted designs, which had to compete with over sixty German, French, and English designs and a dozen American ones. Three remarkable things were disclosed when the judges announced the awards: all of the prize-winners were Americans, all were women, and all worked for Associated Artists. Wheeler was awarded the first prize, Ida Clark second, Caroline Townsend third, and Dora Wheeler fourth.[45]

Popular wallpaper designs of the period, with the exception of those by Associated Artists, consisted primarily of geometric motifs, floral patterns, or figures created in conventional modes. Wheeler expressed the belief that wallpapers should and must be more than this: "To be constantly reminded of the wall as a wall, as a solid piece of

masonry, is what we must avoid . . . and yet . . . it is quite [as] possible . . . to make the wall beautiful as for the pearl-oyster to spread its shell with opalescent nacre." Wheeler's prize winning design was patented on April 11, 1882. The overall design was a silver honeycomb overlaid with a fine network of gold. Superimposed were clover blossoms sprinkled with bees, the clover leaves in gold tints, the tips in brown, the bees in black and yellows. The dado was composed of disks of gold and swarms of bees placed over the clover. The frieze was also gold with the disks simulating beehives placed in regular rows.[46]

Despite their successes with public, private, and commercial commissions, the partnership of Tiffany, Colman, de Forest, and Wheeler was dissolved in 1883 for unknown reasons. Candace Wheeler said the time had come to devote herself "to art in a way which would more particularly help women." This was not possible, she felt, in association with Tiffany, Colman, and de Forest. Because of the great success they had enjoyed, it is possible that all

[44] *Art Amateur* 4, no. 6 (May 1881): 3; *Art Amateur* 5, no. 1 (June 1881): 3.

[45] Wheeler, Document no. 1, n.d., MTMC; M. G. van Rensselaer, "The Competition in Wall-Paper Designs," *The American Architect and Building News* 10, no. 309 (Nov. 26, 1881): 251. Source and information extended to the author by Catherine L. Frangiamore, assistant curator, Cooper-Hewitt Museum.

[46] Wheeler, "Decoration of Walls," *Outlook, A Family Paper* (New York City), Nov. 2, 1895, p. 706; van Rensselaer, "Competition in Wall-Paper Designs," p. 252.

the partners wanted to explore other interests. De Forest commented that the partners had agreed that reorganization of the firm was necessary, yet they could not agree on how to undertake it. Tiffany's increasing interest in glass iridescence has been well documented, and, as Mrs. Wheeler observed, "his wonderful experiments in glass . . . meant far more to him at the time than an association with other interests."[47] Whatever the reasons, the partners separated happily, and Wheeler was allowed to retain the name of the firm.

Associated Artists under Candace Wheeler was primarily concerned with designing fabrics. With her daughter, Dora Wheeler Keith, and Rosina Emmet and Ida Clark, the firm established an American school of design, which specialized in artistic compositions for printed cotton and silk draperies as well as for wallpapers (Fig. 29). The need for fine designs in less expensive textiles, so that every family could share the benefits of attractive fabrics, was of primary concern to Candace Wheeler (Figs. 30–34). The firm studied the manufacture of silk, the printing of cotton, and the design of paper; and it instructed a class of graduates of the Cooper Institute School of Design, to put them in touch with manufacturers. They worked closely with American manufacturers in designing fabrics for an American as well as an international market, giving "chintzes and cottons as much care in the design as . . . expensive brocades."[48]

Although Wheeler was dedicated to making well-designed fabrics available to the lower and middle classes, her reputation and skill brought Andrew Carnegie, Lily Langtry, Mrs. Potter Palmer, Louis Prang, and other wealthy clients to the firm for its services. For Carnegie, the firm designed a damask with the Scotch thistle as the motif, and, for Mrs. Langtry, an elaborate set of bed hangings. From California and from Europe, clients came to Associated Artists to commission original fabrics, hangings, and wallpapers. Further distinction came with the development of "the American tapestry," which Wheeler invented and patented in 1883 (Fig. 35). Appreciating the medium of tapestry for its suggestive qualities, but realizing that the curtain for the Madison Square Theatre had been too realistic, Wheeler wanted to create a tapestry that

THE DESIGNING-ROOM.

FIG. 29. Designing room of Associated Artists. From *Harper's Monthly* 69, no. 411 (Aug. 1884): 345. (Photo, Mark Twain Memorial.)

would be suggestive in a painterly sense, yet distinctive and practical. Because dust and moths would destroy wool, she decided to use silk for the warp and the woof of the canvas. The face of the canvas was covered with embroidery silk, which passed under the slender warp and was actually sewn into the woof. Wheeler called the process "needle weaving," as the needle served as the shuttle carrying the threads. Though expensive to produce, well-to-do clients enthusiastically purchased needle-woven tapestries for their homes.[49]

The variety and scope of Associated Artists' work under Candace Wheeler's direction was widely recognized and well received, and she herself gained the distinction of being America's foremost textile designer. In 1907 the firm of Associated Artists terminated its practice, possibly because Candace

[47] Wheeler, Document no. 1, n.d., MTMC.

[48] Wheeler, "Art Education for Women," in *Outlook,* Jan. 2, 1897, p. 86; Wheeler, Document no. 1, n.d., MTMC; Harrison, "Some Work of the 'Associated Artists,'" p. 345.

[49] Wheeler, "Decoration of Walls," p. 705; "The Art Museum," *Wellesley College Bulletin* 2, no. 1 (Jan. 1936): 9b.

FIG. 30. Associated Artists, fish design on fabric. New York City, 1883–1907. Light blue and white denim; H. 18½″, W. 53″. This design is known to have been manufactured also in dark blue and white denim, green and cream denim, and blue and white cotton. (Collection of the Mark Twain Memorial, gift of Mrs. Francis B. Thurber III.)

FIG. 31. Candace Wheeler, *Consider the Lilies of the Field,* pair of portieres. New York City, 1883–1907. Body of portieres is design of lilies and reeds on muslin, embroidered in outline and seed stitches and with painted highlights; border is made up of 3⅛″ beige cotton sateen with red and green roping, and 13½″ green serge; H. 73″, W. 41½″. (Collection of the Mark Twain Memorial, gift of Mrs. Francis B. Thurber III.)

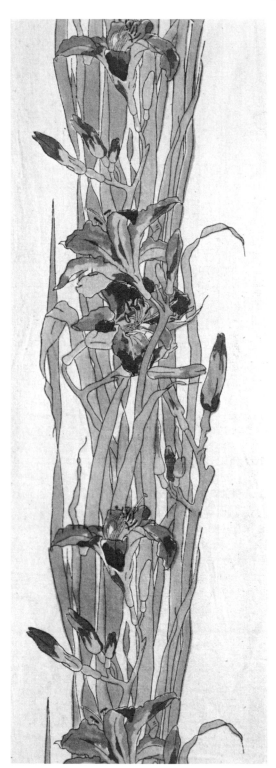

FIG. 32. Associated Artists, design of yellow day lilies on fabric. New York City, 1883–1907. Cotton print; H. 31″, W. 9″. (Collection of the Mark Twain Memorial, gift of Mrs. Francis B. Thurber III.)

FIG. 33. Associated Artists, lily design on fabric. New York City, 1883–1907. Light blue on white cotton; H. 20″, W. 18¼″. Note AA in the design. (Collection of the Mark Twain Memorial, gift of Mrs. Francis B. Thurber III.)

Wheeler had reached eighty years of age. In both periods of its operation, 1879 to 1883 and 1883 to 1907, Associated Artists provided innovative leadership in American design. Walter Smith, in 1876, had found American decorative arts inferior to their European counterparts, but in less than four years the designs of Associated Artists were known internationally. Through reviews in art journals and the writings of Candace Wheeler, the firm's philosophy of room decoration and color treatment was widely read. These American designs by an American firm using principally American materials were, perhaps, Associated Artists' greatest contribution to decorative arts and to American manufacturers. As designers, the partners sought and achieved unique solutions, such as covering a ceiling with seashells, lightening woodwork with metallic patterns, creating popular papers from insects and spiderwebs; and in so doing they encouraged others to consider previously unheard-of possibilities. They replaced heavily lambrequined and shuttered windows with stained glass transoms and fine muslin; they lightened dark walls with pale tints. In general, they challenged the accepted schemes of the past, and they provided the kind of art style that, they felt, should properly represent the tastes of their time. They had set out to do precisely this and, encouraged by clients and manufacturers, made a success of altering the tastes of the upper classes. The work of Associated Artists marked the maturity of an American palette and American designs based on American sensibilities.

FIG. 34. Associated Artists, design of scallop shells with ribbons on fabric. New York City, 1883–1907. Brown outline with yellow on beige, ribbed cotton; H. 64½″, W. 28¼″. Note AA in the ribbon. (Collection of the Mark Twain Memorial, gift of Mrs. Francis B. Thurber III.)

FIG. 35. Candace Wheeler, needle-woven tapestry after the original cartoon by Raphael, *The Miraculous Draught of Fishes*. New York City, 1883–1907. From Candace Wheeler, *The Development of Embroidery in America* (1894; reprint ed., New York: Harper & Bros., 1921), p. 131. (Photo, Mark Twain Memorial.)

# Ruins, Romance, and Reality

## Medievalism in

## Anglo-American Imagination and Taste, 1750–1840

*Alice P. Kenney and Leslie J. Workman*

THE GOTHIC REVIVAL in America is a phenomenon more observed than understood. Why did the future-oriented citizens of Jackson's democracy suddenly turn to building homes, churches, and public buildings in a style clearly suggestive of medieval Europe? This paper will consider the Gothic revival as part of an international trend visible over three centuries in literature and history as well as in art and architecture. It will begin with the definition of "Gothic" and "revival" by Renaissance humanists and touch on the destruction of medieval institutions and the buildings that housed them by religious reformers and political rebels in early modern times. It will consider the paradoxical fusion in England of religious and political reform with the development of a sense of national identity founded in medieval traditions, particularly the common law, so that the very ruins of castles and convents came to be venerated as symbols of the antiquity and continuity of English culture. It will study the derivation from medieval ruins of fashionable decorative motifs for houses, gardens, and furnishings, and of picturesque imagery for paintings, poems, and novels —and the extension of this imagery made possible by the rediscovery of medieval literary romance. It will investigate the reciprocal influence of these fashionable Gothic genres and historical knowledge of the Middle Ages, particularly as fused in the overwhelmingly popular works of Sir Walter Scott, whose effect on his readers' visualization of the medieval past was incalculable.

It should be emphasized that throughout the period under study, "Gothic" was not nearly so carefully defined historically as it is today, being used loosely to characterize various aspects of the thousand years following the collapse of the Roman Empire in the West. This paper will concentrate on the emergence of Gothic visual elements, such as the pointed arch, as common conventions of artistic, literary, and historic communication in England; their transmission to America in books; and the beginning of their appreciation and use by Americans in literature and arts of design. In eighteenth-century England the general public was familiar with the appearance of Gothic architecture because whole buildings as well as ruins surviving from the Middle Ages were fairly common. Many such buildings were still in use. A variety of travelers' descriptions and imaginative interpretations of this type of architecture, including paintings, prints, furniture, poems, and novels, popularized the vocabulary used to describe the impressions that it evoked. Early Americans, having no medieval ruins before their eyes, depended on a few scattered imported illustrations if they wished to interpret literary allusions to the Gothic tradition. As time went on, their understanding was broadened through buildings erected by immigrant architects and descriptions penned by tourists who had seen Gothic ruins and relics in Europe. Gradually, American artists and writers learned to use Gothic elements in their own works.

The conception of "medieval" arose when Italian humanists of the fourteenth and fifteenth centuries used various terms such as "revival" or "awakening" to describe their interest in the Greek and Latin classics. One of these names, "Renaissance," came into use in early sixteenth-century France and was officially sanctioned in the dictionary of the French Academy in 1718. Various terms were used for the period between the Renaissance and the ancient world: Giorgio Vasari referred to the "barbarous Gothic style" of architecture; and *"medium aevum"* was the term that was generally used in the early seventeenth century to refer to this period. The Reformation characterized the Middle Ages as a period of religious, as well as literary and artistic, darkness, a darkness that appeared only deeper in the light of the scientific discoveries of Galileo, Bacon, Leibniz, and Newton. By the end of the seventeenth century, the humanist complaint of bad Latin became a wholesale dismissal of medieval literature, Latin or vernacular; Protestant (or even Catholic) attacks on specific abuses grew into a sweeping condemnation of "monkish ignorance"; the Ptolemaic system was discarded in favor of the Copernican, and the whole of medieval science and philosophy was discarded with it. Since humanist values dominated the educational system of early modern Europe almost without opposition, it is easy to see how terms like "Gothic" or "medieval" could assume a wide variety of meanings, depending on what one wished to condemn, but all of them pejorative and few of them historical as we understand the word.

It is therefore not surprising that in the seventeenth and eighteenth centuries knowledge of medieval history was severely limited except for some antiquarian researches not widely available. Christoph Keller's brief *History of the Middle Ages from the Times of Constantine the Great to the Capture of Constantinople* (1688) achieved a considerable circulation and set limits to the period that are still, to a certain degree, accepted; but the usual opinion was expressed by Voltaire: "It is necessary to know the history of that age only in order to scorn it." [1] In France a sharp break with medieval tradition resulted from the wholesale adoption by Louis XIV and his court of the ideals of absolute monarchy, baroque architecture, and neoclassical standards of literary and artistic criticism, and these values and fashions were accepted by ambitious benevolent despots all over Europe. In Germany an even more drastic cultural hiatus resulted from the political upheavals of the Reformation and the destruction brought about by the Thirty Years War; artistic and literary activities came to

a standstill and did not recover until well into the eighteenth century. By contrast, the continuous development of English thought and culture from medieval to modern times was unique, and it was to England that other European nations turned for a model when they finally became interested in rediscovering their own medieval past.

In England antiquarian research received its first stimulus from religious controversy. Early in the sixteenth century Archbishop Matthew Parker and his associates undertook the study of Anglo-Saxon, since, in the words of a later Saxonist, "by the ancient Saxon monuments we are able to demonstrate that the Faith, Worship and Discipline of our Holy Church is in great measure the same with that of the primitive Saxons." In the seventeenth century the problem of civil rather than doctrinal liberties took precedence and sent lawyers and antiquarians like Edward Coke and John Selden back to the conventional limit of legal memory (1154) and even earlier. Shakespeare's plays, which continued to be read and sometimes performed, expressed a patriotic awareness that the national character had developed in those dark times, but by the eighteenth century this awareness sat rather uncomfortably with fashionable contempt for the Middle Ages. William Blackstone clearly expressed this ambivalence: "That ancient collection of unwritten maxims and customs, which is called the common law, . . . had subsisted immemorially in this kingdom; and, though somewhat altered and impaired by the violence of the times, had in great measure weathered the rude shock of the Norman conquest. This had endeared it to the people in general, as well because its decisions were universally known, as because it was found to be excellently adapted to the genius of the English nation. In the knowledge of this law consisted great part of the learning of those dark ages." [2] Blackstone's *Commentaries* (1765), used even more widely in America than in England as a textbook for lawyers, thus introduced the concept of "those dark ages" to many people who, having no taste for fashionable literature, otherwise might not have been exposed to it.

The events of the Reformation and the civil war left many of England's medieval buildings in ruins.

[1] The authors wish to express appreciation to the Shell Assists program for a faculty development grant received through Cedar Crest College, to the New York State Historical Association for a Dixon Ryan Fox research grant, and to Joseph W. Hammond, Dr. Susan Smith, and Lee Johnson for research assistance. Voltaire quoted in Wallace K. Ferguson, *The Renaissance in Historical Thought* (Cambridge, Mass.: Houghton Mifflin, 1948), p. 89.

[2] John Fortescue-Aland quoted in Arthur Johnson, *Enchanted Ground* (London: Athlone, 1964), p. 220; Blackstone, *Commentaries on the Laws of England. From the Nineteenth London Edition,* 2 vols. (Philadelphia: Lippincott, 1910), 1:15.

The sudden dissolution of the monasteries in 1540 resulted in their desertion, desecration, and even destruction; a century later, the capture and destruction of castles by Cromwell's victorious army rendered them useless as strongholds. Many of these medieval landmarks had been described by Elizabethan antiquarians such as William Camden. Far more significant for their preservation as monuments was William Dugdale's *Monasticon Anglicanum* (1654–73), a comprehensive account of English monastic houses that began to appear during the civil war. This compendium stands as an example of the accomplishments of many seventeenth-century antiquarians who collected curious facts about local history, sometimes suggested by the architectural features of their country houses or revealed in the black letter documents left by their ancestors.[3]

Joseph Addison, early in the eighteenth century, was apparently the first to borrow the term "Gothic" from descriptions of architecture and apply it to literature: "Poets who want the strength of Genius to give that Majestick Simplicity to Nature, which we so much admire in the Works of the Ancients, are forced to hunt after foreign ornaments. I look upon these Writers as *Goths* in Poetry, who, like those in Architecture, not being able to come up to the beautiful Simplicity of the old *Greeks* and *Romans,* have endeavoured to supply its Place with all the Extravagance of an irregular Fancy. . . . the taste of most of our *English* Poets, as well as Readers, is extremely *Gothick.*" Gothic extravagance evidently offended against the elegant simplicity that neoclassical canons of taste characterized as "nature." The conforming of the Gothic to nature, or, rather, the adaption of the neoclassic concept of nature to include the Gothic, took place with the extension to architecture of two ideas from the sizable literature on landscape gardening: first, quite early in the century, the idea that the Gothic in architecture was a direct imitation of nature—carrying the garden, as it were, indoors—and, second, the principle of irregularity, which was developed in 1757 in Edmund Burke's essay, "The Sublime and the Beautiful." It should be noted that these ideas are much more visual than the earlier idea of natural simplicity: the sim-

plicity of Newtonian cosmology is intellectually perceptible; the profusion of the garden is visually perceptible. As Burke observed, "Nature has at last escaped from their discipline and their fetters; and our gardens, if nothing else, declare we begin to feel that mathematical ideas are not the true measure of beauty."[4] In time, this idea was to be the most important element in the developing aesthetic of the "picturesque."

Understanding of these eighteenth-century theories may be impeded by confusion deriving from different meanings then implied by the term "Gothic." Most simply, broadly, and usually it meant "not in the classic style." It also was used in references to architecture that is now called Romanesque, but was then called ancient Gothic. What is now called Gothic was, in the eighteenth century, called Saracenic, Arabic, arabesque, or modern Gothic. Ancient Gothic was condemned as rude, ponderous, and gloomy; modern Gothic as light, frivolous, fanciful, and overornamented; both as lacking simplicity, symmetry, and conformity to nature. To post-Addisonian literary critics, "Gothic" meant "barbarous," "medieval," or "supernatural," meanings that sometimes but not always overlapped. In common speech, the term "Gothic," like the almost synonymous "medieval," simply meant anything of which one disapproved; it was applied to tight lacing, dueling, or other people's opinions. The point of fundamental importance, as the great scholar W. P. Ker pointed out, is that the medieval revival in literature, as well as the Gothic revival in architecture, drew its inspiration from architectural survivals of the Middle Ages rather than from literary survivals (chiefly ballad and romance); the latter did not attract popular attention until after the Gothic revival in design and the Gothic novel were in full career.[5]

When Horace Walpole, who created the Gothic novel, decided to make his home at Strawberry

[3] Kenneth Clark, *The Gothic Revival* (new ed.; Harmondsworth: Penguin, 1964), pp. 13–15; Henry A. Beers, *History of English Romanticism in the Eighteenth Century* (rev. ed.; New York: Holt, 1910), pp. 198–99.

[4] Addison, *Spectator,* May 11, 1711, italics original; Burke quoted in A. O. Lovejoy, "The First Gothic Revival and the Return to Nature," *Modern Language Notes* 47, no. 7 (Nov. 1932): 438.

[5] Lovejoy, "First Gothic Revival," p. 438; A. E. Longueil, "The Word 'Gothic' in Eighteenth Century Criticism," *Modern Language Notes* 38, no. 8 (Dec. 1923): 453–60; Ker, "The Literary Influence of the Middle Ages," in *Cambridge History of English Literature,* ed. A. W. Ward and A. R. Waller, 14 vols. (Cambridge: At the University Press, 1907–16), 10: 245.

FIG. 1. *North Front of Strawberry Hill.* From *A Description of the Villa of Mr. Horace Walpole* (Strawberry Hill, 1784), reprinted in *The Works of Horatio Walpole, Earl of Orford,* ed. Mary Berry, 5 vols. (London: G. G. & J. Robinson, 1798), 2:399.

Hill a showplace of Gothic architecture (Fig. 1), his friend Horace Mann wrote him: "Why will you make [Strawberry Hill] Gothic? I know that is the taste at present but I really am sorry for it." Walpole's conception of Gothic was influenced by his many visits to existing buildings, visits that became elaborate "Gothic pilgrimages," and by his study of antiquarian writers such as William Camden, William Dugdale, and Elias Ashmole, many of whose volumes were profusely illustrated (Fig. 2). Details from these works were freely combined and adopted: thus we learn that in Walpole's library (Fig. 3) "the books are ranged within Gothic arches of pierced work taken from a sidedoor case to the choir in Dugdale's St. Paul's. . . . The chimney-piece is imitated from the tomb of John of Eltham earl of Cornwall, in Westminsterabbey; the stone-work from that of Thomas duke of Clarence, at Canterbury." Of actual medieval literature, apart from some late medieval chronicles collected

for his *Historic Doubts on Richard III,* Walpole knew little and cared less; for example, he dismissed Dante as "extravagant, absurd, disgusting, in short, a Methodist parson in Bedlam." [6]

Walpole had no intention of reviving or restoring Gothic architecture as we understand restoration. His aim was to "unite the grace of Grecian architecture and the irregular lightness and solemnity of Gothic." As Wilmarth S. Lewis has observed, "Walpole and his friends had no idea how Gothic buildings were constructed, nor did they care. What they were after was atmosphere." In his *Description of the Villa of Mr. Horace Walpole*

[6] Mann quoted in R. W. Ketton-Cremer, *Horace Walpole* (New York: Longmans Green, 1940), p. 150; *The Works of Horatio Walpole, Earl of Orford,* ed. Mary Berry, 5 vols. (London: G. G. & J. Robinson, 1798), 3:442; Wilmarth S. Lewis, *Horace Walpole's Library* (Cambridge: At the University Press, 1958), p. 14.

FIG. 2. Wenceslaus Hollar, engraving of arched side doors into choir of Old St. Paul's Cathedral, London. From William Dugdale, *The History of St. Paul's Cathedral* (London, 1658), reprinted in Arthur M. Hind, *Wenceslaus Hollar and His Views of London and Windsor in the Seventeenth Century* (London: John Lane, 1922), pl. 38.

FIG. 3. *Library at Strawberry Hill*. From *A Description of the Villa of Mr. Horace Walpole* (Strawberry Hill, 1784), reprinted in *The Works of Horatio Walpole, Earl of Orford*, ed. Mary Berry, 5 vols. (London: G. G. & J. Robinson, 1798), 3:443.

(1784), Walpole was quite frank and explicit about his purpose:

In truth, I do not mean to make my house so Gothic as to exclude convenience, and modern refinements in luxury. The designs of the inside and outside are strictly ancient, but the decorations are modern. . . . But I do not mean to defend by argument a small capricious house. It was built to please my own taste, and in some degree realize my own visions. . . . could I describe the gay but tranquil scene where it stands, and add the beauty of the landscape to the romantic cast of the mansion . . . at least the prospect would recall the good humour of those who might be disposed to condemn the fantastic fabric, and think it a very proper habitation of, as it was the scene that inspired, the author of the Castle of Otranto.

Modern research has indeed shown the action of *The Castle of Otranto* to be very precisely located at Strawberry Hill: "The gallery, round tower and the great cloister at Strawberry—were all built before the publication of this novel, and they are incorporated in the tale." [7]

Walpole's writings made him the founder of a fashion that spread widely and produced other notable edifices, such as William Beckford's Fonthill Abbey—in its turn the setting of Beckford's Saracen-Gothic novel *Vathek*. The *Description of the Villa* is studded with adjectives indicating the emotional effects Walpole intended to create: "You first enter by a small gloomy hall." In the tribune, or cabinet, "the roof, which is taken from the chapter-house at York, is terminated by a star of yellow glass that throws a rich gloom over all the room, and with the pointed windows gives it the solemn air of a rich chapel" (Fig. 4). After this we may be surprised to learn that in writing *The Castle of Otranto*, "that great master of nature[,] Shakespeare, was the model [Walpole] copied," but we will not be surprised at the essentially visual effect Walpole achieved. Actually Walpole's homage to Shakespeare is not as farfetched as it might at first appear. A modern critic has pointed out that the action and themes of the Gothic novel were Renaissance rather than medieval. The emphasis on blood and violence, the revenge motive, the supernatural, poisoning as one of the fine arts,

FIG. 4. *The Cabinet.* From *A Description of the Villa of Mr. Horace Walpole* (Strawberry Hill, 1784), reprinted in *The Works of Horatio Walpole, Earl of Orford*, ed. Mary Berry, 5 vols. (London: G. G. & J. Robinson, 1798), 2:470.

the fondness for Italian settings—all of these belong to the Renaissance tradition of dramatists such as Cyril Tourneur, John Ford, and John Webster, and to the Shakespeare of the tragedies, to whom Mrs. Radcliffe explicitly acknowledged a considerable debt. The style of the Gothic novel is furthermore strongly dramatic, and many passages read like scripts, creating a visual effect similar to that evoked by reading plays. In these dramas it is the setting rather than the action that is "Gothic." [8]

Walpole's passion for "Gothick" decoration carried to an extreme a fashion already popular among his contemporaries. In 1762, two years before the publication of *The Castle of Otranto*, the third edition of Thomas Chippendale's *Gentleman and Cabinet-Maker's Director* appeared, containing more than a score of Gothic designs for library and bedroom, as well as for occasional furnishings,

[7] Horace Walpole, *Correspondence*, ed. Wilmarth S. Lewis, 27 vols. to date (New Haven: Yale University Press, 1937–), 21:250; Berry, *Works of Walpole* 3:397–98; K. H. Mehrotra, *Horace Walpole and the English Novel* (New York: Russell & Russell, 1970), pp. 17–18.

[8] Berry, *Works of Walpole* 3:401, 470; Lewis, *Walpole's Library*, p. 39; Clara F. McIntyre, "Were the 'Gothic' Novels Gothic?" *Publications of the Modern Language Association* 36, no. 4 (Dec. 1921): 664–67.

including two cases for chamber organs. In structure, Chippendale's Gothic designs differed little from his other models, their Gothic character being supplied by carved surface decoration derived from such architectural features as pointed arches, pinnacles, and tracery. Although Chippendale's classical and rococo motifs were far more popular in America, a tea table made in New York between the publication of Chippendale's third edition and the Revolution shows that a few colonists were aware of the new Gothic fashion and were by no means backward in adopting it (Fig. 5). The mahogany table is not a direct copy of any one of Chippendale's patterns, but a composite of elements from two or three.[9] Its foliated cabriole legs are sturdier than Chippendale's usual designs, but overall the table shows far greater simplicity, though no less delicacy in its gallery of Gothic arches, than other Chippendale patterns.

In the same decade an American, Benjamin West, made a contribution to the art of history painting that made it much easier for the English to visualize their medieval history as a heroic past. Soon after West migrated from Pennsylvania to London, in the 1760s, he began to depict historical characters in the costumes of their own time rather than in those of classical antiquity, justifying the innovation as follows: "The same truth which gives law to the historian should rule the painter. If instead of the facts of the action I introduce fictions, how shall I be understood by posterity? I want to mark the place, the time, and the people, and to do this I must abide by the truth."[10] The sensational success of *The Death of General Wolfe* established both his reputation and his theory, which he soon applied to a medieval subject in *The Death of the Chevalier Bayard*. His later copious production included illustrations based on the works of Shake-

FIG. 5. Tea table, New York City, 1765–77. Mahogany; H. 28⅜″, W. 32″, D. 17½″. (Winterthur 52.18.)

speare, Spenser, and Ariosto, scenes based on the legends of King Alfred, and three portraits of medieval religious figures for William Beckford's gallery at Fonthill Abbey. West's most ambitious venture into the Middle Ages was his series of eight paintings for Windsor Castle depicting events in the life of Edward III. *Queen Philippa Interceding for the Lives of the Burghers of Calais* (Fig. 6) emphasizes the dramatic contrast between the violent attitude of the king, who is about to assault the bound and helpless burghers, and the compassion of his beautiful queen, who is pleading for their lives. The painting elevates the politic mercy of Froissart's Middle Ages to the moral level of a patriotic fairy tale. Although recent research has demonstrated that West's versions of fourteenth-century costume and setting were not entirely accurate, he assembled available facts according to familiar iconographic and dramatic conventions and awarded England's own history the significance previously reserved for classical antiquity. In his *Columbiad*, Joel Barlow called American attention to West's accomplishment, referring particularly to this painting:

> West with his own great soul the canvass warms,
> Creates, inspires, impassions human forms,
> Spurns critic rules, and seizing safe the heart,
> Breaks down the former frightful bounds of Art. . . .
> Edward in arms to frowning combat moves,
> Or, won to pity by the queen he loves,
> Spares the devoted *Six*, whose deathless deed
> Preserves the town his vengeance doom'd to bleed.[11]

[9] Chippendale, *The Gentleman and Cabinet-Maker's Director* (3d ed., 1762; reprint ed., New York: Dover, 1966); Ralph Fastnedge, *English Furniture Styles, 1500–1830* (Harmondsworth: Penguin, 1955), p. 166; Joseph Downs, *American Furniture: Queen Anne and Chippendale Periods* (New York: Macmillan, 1952), no. 315.

[10] West quoted in C. Edwards Lester, *Artists of America* (New York: Baker & Scribners, 1846), p. 93. For general information on West, see James T. Flexner, *Light of Distant Skies: American Painting, 1760–1835* (1954; reprint ed., New York: Dover, 1969), pp. 37, 128–29; Henry E. Jackson, *Benjamin West: His Life and Work* (Philadelphia: Winston, 1900), pp. 85–87; Grose Evans, *Benjamin West and the Taste of His Times* (Carbondale: Southern Illinois University Press, 1959), pp. 69–70, 122, pls. 50–52.

[11] Barlow, *Columbiad* (Philadelphia: Fry & Kammerer, 1807), p. 310.

FIG. 6. Benjamin West, *Queen Philippa Interceding for the Lives of the Burghers of Calais*. England, 1788. Oil on canvas; H. 39½″, W. 52¼″. (Detroit Institute of Arts, gift of James E. Scripps.)

A view of history much like West's became popular in the 1770s and 1780s in novels and biographies—such as Alexander Bicknell's *History of Edward Prince of Wales* (the Black Prince, son of Edward III)—that dealt with heroic figures from England's past. The Gothic novels of this period owed more to *The Castle of Otranto* than to historical research, since their historical content consisted of the appearance of some famous personage, such as Henry II or Queen Elizabeth, who participated in an entirely fictitious series of events. These characters were readily identifiable by readers whose knowledge of history was rudimentary; the stories were shaped by popular demand and the imagination of novelists, many of them women, who were but slightly acquainted with historical research even as it then existed. In fact, as J. M. S.

Tompkins has pointed out, novelists derived their sense of the past far more from contemplation of ruins than from knowledge of historical facts. "Romance was to find the materials for its new appearance in the past, and in a past which, though sometimes nominally historic, is really an elaboration of the impressions made by Gothic architecture on modern sensibility. . . . It must be premised that the historical novel as we understand it, that union of exact knowledge and penetrating imagination, was not at this time possible; the historic sense was too little developed." [12] This attitude came to the

[12] Tompkins, *Popular Novel in England, 1770–1800* (Lincoln: University of Nebraska, 1961), pp. 227–28, 234. See also Warren H. Smith, *Architecture in English Fiction* (New Haven: Yale University Press, 1934), pp. 107–21, 131–32, 137.

fore in the 1790s, when historical novels went out of fashion and architectural description increased sharply in thrilling tales set vaguely in castles or convents, like Mrs. Radcliffe's best-selling *Mysteries of Udolpho* and Matthew G. Lewis's sensational *The Monk*.

Another literary phenomenon of the late eighteenth century falls within the broad limits of the medieval revival: the Welsh and Norse poems of Walpole's friend Thomas Gray, the Ossianic poems of James Macpherson, and the Rowley poems of Thomas Chatterton. These works, like the Gothic novel, had very little to do with the historical Middle Ages; although Gray did translate some Welsh and Norse verse in addition to writing original poems on Welsh themes, the poems Macpherson attributed to Ossian were almost entirely original compositions inspired by Gaelic fragments, and Chatterton's poems were original compositions suggested by physical remains—a church with its black letter Bible and dusty muniment room. It is significant that even when Macpherson's and Chatterton's "discoveries" were unmasked as impostures, their popularity was little diminished. Gray, for example, suspected the authenticity of the Ossianic poems from the first but championed the work in its own right. Here the eighteenth century was essentially expressing itself; what it found in Ossian and in Gray's poems was the "sublime." Of Gray's translations from the Norse, *The Descent of Odin* and *The Fatal Sisters,* Ker remarked very significantly that they "exactly correspond to his own ideals of poetic style. Gray must have felt this. . . . it was all too finished, too classical. No modern artist could hope to improve upon the style of the northern poems." [13] This, Ker concluded, is why (apart from his habitual indolence) Gray pursued this interest no further.

Nevertheless, except for a few minor imitators of Gray, eighteenth-century authors disregarded the potential for "sublimity" of the great quantity of Welsh history and literature that became available in that century in both Welsh and English, and virtually nothing was done either to publish or otherwise to exploit the incomparably larger body of Irish material. Scotland fared better, even before the time of Sir Walter Scott, thanks to the work done by patriotic antiquarians unwilling to extend the Act of Union to culture. Virtually as much his-

torical information about the British Middle Ages was available at the beginning of the eighteenth century as was accessible to Scott at the beginning of the nineteenth century; what was lacking in the earlier period was the motivation to exploit these materials. Seventeenth-century antiquarian erudition declined sharply in the first quarter of the eighteenth century: the work of men like Camden and Dugdale and their successors had been produced amid religious and constitutional crises, and, when these were resolved, the public, and perhaps also the incentive, for such work disappeared. Furthermore, in the new religious climate of Deism, the patronage of the Anglican church, which had been of great importance, was withdrawn, and private patronage declined by reason of the Augustan contempt for antiquarians. For example, John Pinkerton, in the preface to his *Ancient Scotish Poems,* expressed the hope that "the reader will allow . . . that the editor has in no instance sacrificed the character of a man of taste to that of an antiquary; as of all characters he should the least chuse that of a hoarder of ancient dirt." [14]

This attitude went so far that historians who used the antiquarians felt obliged to apologize for the fact—even Edward Gibbon, whose scholarship stands out amid such eighteenth-century historians as David Hume and William Robertson. The medieval volumes of Hume's *History of England* were undertaken only at his publisher's insistence, for the sake of completeness, after the later period had been finished. Hume, whose indifference to available medieval authorities was remarked even in his own time, promised in the opening paragraph that "we shall hasten through the obscure and uninteresting period of Saxon annals; and shall reserve a more full narration for those times when the truth is both so well ascertained and so complete as to promise entertainment and instruction to the reader." [15] Nevertheless, the researches of the antiquarians were being communicated to the eighteenth-century reading public. The *Gentleman's Magazine* published a steady stream of antiquarian articles in the last decades of the century, and there was a steady increase in the patronage and publication of local histories. The gap between primary

[13] Ker, "Literary Influence of the Middle Ages," p. 254.

[14] Pinkerton quoted in Keith Stewart, "Ancient History as Poetry in the Eighteenth Century," *Journal of the History of Ideas* 29, no. 3 (June 1958): 337–38.

[15] Hume, *History of England,* 6 vols. (1754–61; reprint ed., New York: Harpers, 1879), 1:2.

research and general knowledge was being closed, and a public was being created for historical writing, which would lead in time to a truer sense of the past.

This development is more apparent in the areas of literature and literary scholarship. Gray, who had devoted considerable research to his Welsh and Norse poems, had shown the way, but the real marriage of literature and scholarship took place in the genres of ballad and romance, two medieval literary forms that had never quite died out and which were consequently suited to form a bridge between eighteenth-century Gothic and the Middle Ages. Romance had ceased to be written in England after Spenser, but Spenser continued to be read by authors and gentlemen of literary taste. Addison's famous defense of *Chevy Chase* in the *Spectator*—on the strictest classical grounds—indicates the strength of the ballad, which survived in chapbooks as well as in popular memory. Bishop Thomas Percy, usually regarded as England's pioneer ballad collector, was of course less of an innovator than has been popularly supposed. There had been previous ballad collections, but Percy's was better and was supported by a scholarship tastefully selected from the despised antiquarians. Another literary scholar who achieved wide popularity was Richard Hurd, whose *Letters on Chivalry and Romance*, based on the *Mémoires sur l'ancienne chevalerie* by Saint-Palaye, did much to change opinion in favor of the Middle Ages. Thomas Tyrwhitt's knowledge of Middle English enabled him to expose Chatterton's Rowley poems as forgeries and, even more significantly, to reopen the whole world of Middle English verse to modern readers by recovering the long-lost understanding of Chaucer's metrics.

Thomas Warton's three-volume *History of English Poetry* (1774–81), the triumphant completion of an undertaking Pope had projected and Gray had attempted, provided a picture hitherto entirely lacking of medieval intellectual history and first made possible consideration of the development of language and literature. Warton combined social, literary, philosophical, and religious elements to create a picture of the Middle Ages in sharp contrast to the ideals of eighteenth-century classicism, a picture that, though not very far removed from the Dark Ages of the humanists, was consistent and positive, and was derived from the known sources. Warton's work provides a convenient summation of the understanding of the time.

The tournaments . . . of our ancient princes . . . while they inculcated the most liberal sentiments of honour and heroism, undoubtedly contributed to introduce ideas of courtesy and to encourage decorum. Yet the national manners still retained . . . a mixture of barbarism, which rendered them ridiculous. This absurdity will always appear when men are so far civilized as to have lost their native simplicity, and have not yet attained just ideas of politeness and propriety. Their luxury was inelegant, their pleasures indelicate, their pomp cumbersome and unwieldy. In the mean time it may seem surprising, that the many schools of philosophy which flourished in the middle ages, should not have corrected and polished the time. But as their religion was corrupted by superstition, so their philosophy degenerated into sophistry.[16]

The twenty years from the publication of Warton's *History of English Poetry* (1774–81) to that of Scott's *Minstrelsy of the Scottish Border* (1802–3) saw a steady growth in the medieval revival, with a few significant changes in emphasis. Of these, perhaps the most important was the growth in the popular appeal and in the number of literary descriptions of the picturesque. By the turn of the century there was a considerable literature of guides to picturesque scenery, particularly in the Lake District, a consequence of the improved accessibility of these areas to tourists and the increasing popularity of travel as recreation. Uvedale Price's *Essay on the Picturesque* (1794) added a third category to the sublime and the beautiful and "set up irregularity in place of regularity as the essential of design" in architecture.[17] These theories were further supported by the revolutionary principle of aesthetics proclaimed by Archibald Alison in his *Essay on the Nature and Principles of Taste* (1790), in which he attributed feelings evoked by landscapes to associations in the mind of the spectator rather than to qualities inherent in the object of contemplation.

Simultaneously a popular craze for archaeology, reflected in the pages of the *Gentleman's Magazine*, led to a great increase in the publication of local histories. But local histories remained expensive and consequently of limited impact until the publication of John Britton's *Beauties of Wiltshire* (1801) demonstrated the existence of a large popular market for cheaper works of this kind, a market

[16] Warton, *History of English Poetry*, 3 vols. (1774–81; reprint ed., London: Ward, Lock, 1870), p. 224.

[17] Price quoted in Christopher Hussey, *The Picturesque: Studies in a Point of View* (London: Putnam, 1927), p. 187.

Britton hastened to supply. One effect, as Kenneth Clark observed, was that "Britton killed Ruins and Rococo" [18]—the fantastic Gothic decoration of the eighteenth century—by calling attention to authentic Gothic structures from the Middle Ages. The antiquarian Joseph Strutt, originally an engraver, published a series of highly popular works on medieval manners and customs (1775–89). Sharon Turner's *History of the Anglo-Saxons* (1799–1805), based on extensive research in Saxon and British sources, marked a new departure in this neglected area, although his indifferent style prevented his securing the audience enjoyed by Hume or Gibbon. Furthermore, Turner's work was immediately overtaken by John Lingard's *The History and Antiquities of the Anglo-Saxon Church* (1806 and 1810), a much shorter and more readable work—a painful exercise in objectivity by a Catholic historian and one that helped to precipitate the revival, in the next generation, of medieval religious customs.

Had Walter Scott, like Keats, died at twenty-six, he would be barely remembered as the author of some minor though pioneering translations from German romantic poetry and some unimportant ballad imitations. At thirty-one he was the author of *The Minstrelsy of the Scottish Border,* a pioneering work of ballad collection supported by an impressive apparatus of antiquarian lore. At thirty-six, the age at which Byron died, Scott was the author of two long narrative poems, *The Lay of the Last Minstrel* and *Marmion,* which made him Great Britain's most famous poet and perhaps the best-known Briton of his day. In 1810 *The Lady of the Lake* shattered all previous sales records for poetry; and in 1814 *Waverley* appeared, the first in a series of twenty-six major works in the genre that it initiated, the historical novel. These achievements were the work of a busy public official actively engaged in journalism, editing, publishing, antiquarian research, an enormous literary correspondence, and the active patronage of other writers. They were the product of great energy, determination, and discipline, aided by a prodigious memory. But it is clear that while such qualities might account for the amazing quantity of Scott's work, they could not alone account for its consistently high quality. Still less could they sufficiently explain Scott's equally astonishing popularity, which remained as high throughout the nineteenth century as it was during his lifetime, although it has declined in the twentieth century.

Scott was peculiarly fortunate in both his time and his place. The realistic novel, already a powerful social force, was ready to extend its range, and the Gothic novel might sustain a dimension of reality. The development of historical sources has been mentioned, and it is fair to say that, by Scott's time, historians writing in Great Britain had mastered the subject matter of the Middle Ages, but not the art of telling a story. "The difference between Gibbon and Macaulay," wrote George M. Trevelyan, "is a measure of the influence of Scott." Edinburgh in 1800, no longer the capital of an independent nation but still the cultural capital of a stubbornly independent people, may not have been quite the self-styled Athens of the North that Thomas Love Peacock ridiculed, but it was a lively intellectual center. Equally important, it was also the bustling commercial center of a country whose rural society was still feudal but whose economy had been transformed by its incorporation into the trading network of the British Empire. As a result, the cultural frontier was no longer the Tweed but the Highland line, and a dramatic contrast of manners was observable between medieval countrymen and modern city-dwellers, as well as between anglicized Lowland Scots and still-Gaelic Highlanders. These contrasts were the theme of *Waverley* and ran throughout the Waverley novels. As Scott explained, "The ancient traditions and high spirit of a people who, living in a civilized age and country, retained so strong a tincture of manners belonging to an early period of society must afford a subject favorable for romance." [19]

Raised in the border country and very conscious of his own ancient family, Scott was ideally suited to bring ballad research out of the library and into the countryside; for him the modern ballad was not an imitation but a living tradition. Furthermore, Scott the lawyer was well aware of the traditional continuity of the common law, which he compared to continuity in human behavior and in literature. As one of his characters remarked, "A lawyer without history or literature is a mechanic, a mere working mason; if he possesses some knowledge of these he may venture to call himself an architect." In this sense, Scott's own life and work are inseparable; lawyer, magistrate, businessman,

---

[18] Clark, *Gothic Revival,* p. 30.

[19] Sir Walter Scott, *Waverley Novels,* 48 vols. (London: J. C. Nimmo, 1893), 1:xx.

and lord of the manor, a self-made man of ancient family, easy and intimate with people in all walks of life, Scott was singularly equipped by nature and circumstance to bring a new realism to both history and fiction. Edgar Johnson, his definitive biographer, has spoken of his "massive sanity." [20] In England, only Chaucer and Shakespeare have similarly reflected the whole range of their times.

Scott was as interested in places as he was in people. He took pains with the local and architectural background of his poems and novels, engaging in extensive correspondence if he could not visit a place himself at a certain season to observe the vegetation. In 1831, the year before his death, he insisted on visiting Lanarkshire to verify the locale of *Castle Dangerous,* on which he was then working, "even though the topography as such plays no part in the plot and is barely mentioned in the book." Scott's descriptions, rather than Macpherson's, made the Highlands a tourist attraction; every new poem or novel added to the sites to be visited. A recent critic has suggested that Scott's wide reading of Gothic novels is what probably led to the importance of architecture in his fiction. But Scott's early antiquarian interests, quite apart from his childhood in the border country, where every ruined peel tower had its story, can more than account for it. Nor can we agree that his architecture, "in picturesque qualities and emotional effect, is far inferior to the architecture of Mrs. Radcliffe." [21] Although Scott's architecture stays in the background, where it belongs, the realism and accuracy of his descriptions, and the part the buildings themselves play in the action, give a new dimension to fiction.

Scott's castellated residence at Abbotsford (Fig. 7) reflected his keen sense of the importance of fitting surroundings for the roles he was called on to play in the essentially feudal society of his county. The critics who considered his expenditures on Abbotsford extravagant have perhaps given insufficient recognition to his inherited and acquired social position and the values and expectations of his contemporaries. When undertaken, the rebuilding of Abbotsford was entirely justified both by the unprecedented fortune Scott had made from his writings and by the position he occupied as laird of a large estate, representative of an ancient family, and sheriff of the county; the financial debacle that made Abbotsford a burden to him was a result of the undermining of traditional institutions by modern economic currents that were understood by no one at his time. Certainly there is a great difference between Abbotsford and Strawberry Hill or Fonthill Abbey. Abbotsford was not a gimcrack or a folly, but a solid house, belonging to its countryside and built to be lived in. If it looked forward to the extravagance of the Scottish baronial style, it also looked back to the old border keep. This relationship between the house and the architectural traditions of its locality was no accident; all his life Scott was actively interested in the preservation of ancient buildings, particularly Melrose Abbey (Fig. 8), which he expected every visitor to Abbotsford to see. It has been suggested that "what Scott [did]— and for the first time in either fiction *or* history— [was] to dramatize the basic processes of history." [22] He did this by developing his characters in the context of historical circumstances and setting them off against a fully realized physical background. Just as the romantic setting in Scott's fiction was a necessary part of the development of the story, and not just a backdrop, so the medieval decoration of Abbotsford was not a trapping, but an inherent part of his way of life.

Scott's novels, which began to appear just as the Napoleonic wars were ending, immediately became as popular on the Continent as in England, bringing a common element to an otherwise varied and confusing European culture. In the wake of the German romantic movement in literature and philosophy, the revolutionary spirit generated by Napoleonic oppression stimulated great advances in the study of history, philology, and the study of politics. In France, Revolutionary and Napoleonic excesses dimmed the luster of reason as the intellectual road to the millennium, and the leading literary figure was François René de Chateaubriand, whose *La Génie du christianisme* (1802) offered consolation in the form of a romanticized Catholic Middle Ages. The German and French movements, though contradictory in their political implications, shared a common interest in the Middle Ages

[20] G. M. Trevelyan, "Influence of Sir Walter Scott on History," *Autobiography and Other Essays* (London: Longmans Green, 1949), p. 201; Scott, *Waverley Novels: Guy Mannering* 2:421; Johnson, *Sir Walter Scott: The Great Unknown,* 2 vols. (New York: Macmillan, 1970), 2:1254.

[21] J. H. Paterson, "The Novelist and His Region: Scotland through the Eyes of Sir Walter Scott," *Scottish Geographical Magazine* 81, no. 4 (Dec. 1965): 146–47; Smith, *Architecture in English Fiction,* p. 176.

[22] Johnson, *Sir Walter Scott* 2:1290.

FIG. 7. James Johnstone, *Abbotsford*. Edinburgh, 1833. Engraving; H. 6⅞″, W. 9 9/16″. (Prints Division, New York Public Library.)

as the source of their respective ideals of liberty and extreme conservatism, and in the next few years there emerged a powerful school of romantic historiography. In Germany it was led by Leopold von Ranke, the founder of the modern historical method, and in France by medievalist Augustin Thierry, conservative statesman François Guizot, and, most popular of all, Jules Michelet. Thierry explicitly acknowledged the indebtedness of medieval historians to Scott: "The reading of the romances of Walter Scott has turned many thoughts toward the Middle Ages from which not long ago one turned away in disdain; and if in our time there should be a revolution in . . . history these works will have contributed to it in a singular way." [23] The corresponding impulse in England

focused attention on the tradition of parliamentary liberty that had sustained the struggle against Napoleonic tyranny to the triumph of Waterloo, so that Scott's lessons in historical writing were first turned to advantage by Thomas Babington Macaulay in his *History of England* (1849–61), which began with the antecedents of the Glorious Revolution of 1688.

In the meantime, medieval scholarship in England continued largely in the antiquarian tradition. Henry Hallam's *View of the State of Europe during the Middle Ages* (1818) reflected the renewed interest in the period, but, published only four years after *Waverley* and three years before *Ivanhoe*, his work had little opportunity to profit from the lessons of Scott. A movement was already under way, however, to collect in a single repository official documents preserved in various government departments; this movement led to the estab-

[23] Thierry translated and quoted in J. Westfall Thompson, *History of Historical Writing*, 2 vols. (New York: Macmillan, 1942), 2:229.

Fig. 8. J. B. Mould, *Melrose Abbey*. Edinburgh, 1841. Engraving; H. 2½", W. 4 3/16". (Local History and Genealogy Division, New York Public Library.)

lishment in 1833, a year after Scott's death, of the Public Record Office. It was soon discovered that the completeness and continuity of these archives, which began with the Domesday Book, made them uniquely valuable for the study of medieval economic and social life as well as constitutional development. The eventual exploitation of these materials by the great English medievalists, led by T. H. Round, William Stubbs, and Frederic W. Maitland, lies beyond the scope of this paper. One immediate consequence of the removal of the records to new quarters was the destruction of the Houses of Parliament by a fire ignited to burn discarded Exchequer tallies, and the subsequent decision to rebuild that symbolic edifice in the Gothic style.[24]

This spectacular event established Gothic architecture as a symbol of English nationality at a time when London was the Mother of Parliaments, the world's financial center, and the capital of the largest empire the world had ever seen. At the same time, reformers hoping to make the Anglican church more truly national were seeking ways for it

to reach a greatly expanded and redistributed industrial population and to meet the neglected needs of some churchgoers for emotional and aesthetic experience in worship. The former purpose prompted the Church Building Act of 1818, under which 214 new churches were erected, 174 of them "in a style then described as Gothic";[25] the latter prompted the Oxford movement and the closely related ecclesiological revival, which directed antiquarian research toward liturgical practices and their architectural setting. These patriotic and religious movements shared a driving force of moral earnestness to revive traditional virtues for the regeneration of industrial England, a motive that caused a number of authors to envision an ideal preindustrial Middle Ages to whose way of life these virtues were fundamental. Such a medieval golden age was first depicted by A. W. N. Pugin, one of the architects of the new Houses of Parliament, in *Contrasts* (1836); but his ideas were suspect because he could not separate them from his Catholic faith, and it was left for Thomas Carlyle to catch popular attention with a similar theory in

[24] Clark, *Gothic Revival,* p. 93.

[25] Clark, *Gothic Revival,* p. 81.

*Past and Present* (1843). In the same year John Ruskin, in *Modern Painters,* advocated moral earnestness as a criterion of art criticism; his friends of the Pre-Raphaelite Brotherhood, founded in 1848, applied this principle to their depiction of medieval subjects and their study of medieval craftsmanship, thereby turning the Gothic revival in England into an entirely new channel.

The transplantation to America of this movement in all its variety was clearly impossible, for the history of American communities neither dated from the Middle Ages nor contained the "dates of battles and deaths of kings," of which medieval history then consisted. There were no ruins in America nor relics save those of the Indians; and American writers had neither the techniques for making use of the surviving medieval elements in the folklore, particularly the ballads, of the people around them nor the backgrounds against which to set medieval tales of their own invention. Furthermore, the highly visual nature of medievalism severely hampered its transmission, particularly when the American Revolution, occurring at the height of the Gothic fashion, interrupted communication between England and America. For the remainder of the eighteenth century, Americans were perforce limited to what they could learn of medievalism from a scattering of the books available to English readers. As Constance Rourke has observed, "The effects of literacy were . . . magnified by isolation with the result that dependence upon the printed word early became a common social trait." [26]

English books read in America included the histories by Hume, Gibbon, and Robertson. American revolutionary publicists usually did not trouble to trace the descent of Anglo-American liberty further back than John Locke, so that the English revolutionary emphasis on Magna Charta as the fount of liberty dropped into the background. The great antiquarian compilations were too bulky, too expensive, and too localized to cross the Atlantic in any quantity, as were sumptuous books of views and the picturesque landscapes of the English watercolorists. Americans' visualization of the elements of Gothic design was limited to the contents of such craftsmen's guides as those by Batty Langley and Thomas Chippendale, whose patterns in

this style were not widely executed in eighteenth-century America. *The Castle of Otranto* and the folklore collections found few American readers and no imitators at this time. The Ossianic poems were sufficiently popular to justify allusions and parodies by a number of the first generation of American authors, including John Trumbull, Joel Barlow, and Charles Brockden Brown. Gothic tales of terror of the 1790s, unlike the historical novels of the 1770s and 1780s, were eagerly imported; and cultivated gentlemen, especially in Philadelphia and New York, were acquainted with the romances by Spenser, Cervantes, and Ariosto, at least in translation.[27] The next generation of American readers devoured the works of Scott, whose poems circulated widely in spite of the international tension at the time of their publication, and whose novels revolutionized American publishing. There are few reliable statistics about the publishing business during this period; but, while Scott's English publishers considered the sale of 12,000 three-volume copies of one of his novels as surpassing all previous records, it has been estimated that 5 million cheap reprints of the Waverley series alone poured from American presses, and presumably were sold, between 1813 and 1823.[28] Unlike many books whose best market was in the East, Scott's sold in the West as well, evidently providing settlers there with a convenient, portable means of maintaining contact with the culture they had left behind. This wide distribution and the limited variety of books and other cultural resources available in America made Scott's impact there in some respects greater and more lasting than in Europe.[29]

Anglo-American relations improved and contacts increased during the 1790s, partly in reaction to the excesses of the Reign of Terror and partly after a number of disputes with England were resolved by Jay's Treaty. During the undeclared war with France provoked by the XYZ affair, America's first Gothic revival, embracing both architecture and literature, began in Philadelphia, which was then

[26] Rourke, *Roots of American Culture* (New York: Harcourt, 1942), pp. 54–55.

[27] Van Wyck Brooks, *World of Washington Irving* (Philadelphia: Blakiston, 1944), p. 32; Carl and Jessica Bridenbaugh, *Rebels and Gentlemen* (New York: Oxford, 1962), 86–95.
[28] Merle Curti, *Growth of American Thought* (3d ed.; New York: Harper & Row, 1964), pp. 236–38.
[29] Rollin G. Osterweis, *Romanticism and Nationalism in the Old South* (Baton Rouge: University of Louisiana, 1967), p. 41.

FIG. 9. Benjamin H. Latrobe, *Kirkstall Abbey.* Leeds, 1798. Watercolor; H. 5⅛″, W. 8⅜″. From Latrobe, *An Essay on Landscape Explained in Tinted Drawings, 3* vols. (Richmond, Va., 1798), 1:28. (Virginia State Library.)

the nation's cultural and financial capital and would remain so for another generation, although the seat of government was about to be moved to Washington. The architectural leader of this movement was the city's—and the nation's—first professional architect, Benjamin H. Latrobe, who was born in England and was educated there and in Germany. Latrobe's painting of Kirkstall Abbey (Fig. 9), in Yorkshire near his boyhood home, illustrates his early interest in picturesque Gothic and also shows elements of mass and detail that were to reappear frequently in his Gothic buildings. Latrobe saw German Gothic buildings while at school on the Continent and became familiar with Gothic fashion during his apprenticeship to the English architect Samuel P. Cockerell. The plans for Sedgley (1799), the first house of his design in Philadelphia (Fig. 10), called for pointed windows and a crenellated cornice superimposed on a Georgian mass with Tudor decorative details, but Latrobe was bitterly disappointed by the inability of Philadelphia craftsmen to execute the Gothic features of this design. Five years later he submitted alterna-

tive Roman and Gothic plans for the Baltimore cathedral; although the classical plan was erected, the Gothic is worth noting because its exterior was clearly inspired by Kirkstall Abbey while its interior decoration was modeled on German Gothic churches.[30]

Latrobe experimented with Gothic in a major commercial building, the Bank of Philadelphia (1805), for which he adapted familiar motifs from Gothic castles and churches. The rectangular building, with its setback, crenellated third story and its ornamentation of a deep ogival portico, rose window, and clerestory, looked rather like a sawed-off cathedral and bore no resemblance to authentic medieval civic buildings, such as the cloth halls of Flanders. Within the next five years one of Latrobe's pupils, William Strickland, used the Gothic revival style for the Masonic Hall on Chestnut Street (Fig. 11). This structure combined a battlemented castle for the hall and a cathedral spire for the tower, in

[30] Talbot Hamlin, *Benjamin Henry Latrobe* (New York: Oxford, 1955), pp. 6–7, 27, 150–51, 244–48.

FIG. 10. William Birch, *Sedgley the Seat of Mr. Wm. Crammond* (detail). Springland, Pa., 1799. Engraving; H. 5¾″, W. 7¼″. (Historical Society of Pennsylvania.)

this case by no means inappropriate to the building's function. Masons in general were soon informed about the peculiar fitness of this style for their society by an article in the *Free Mason's Magazine* in 1811, and a few years later similar temples were erected in Boston and New York. It is thought that another of Latrobe's pupils, Robert Mills, helped to design the Dorsey house, within a few blocks of the Masonic Hall, and the subject of another magazine article in 1811. These two articles represent the only discussions of Gothic architecture in American periodicals before 1815, with the exception of two descriptions of historic English abbeys by preembargo tourists.[31] All three of these build-

FIG. 11. William Kneas, *Masonic Hall* (detail). Philadelphia, 1813. Engraving; H. 22½″, W. 19⅛″. (Historical Society of Pennsylvania.)

[31] Agnes A. Gilchrist, *William Strickland* (Philadelphia: University of Pennsylvania, 1950), pp. 2, 34, 45–46, pls. 5, 11; "On Gothic Architecture," *Free Masons Magazine and General Miscellany* 2, no. 1 (Oct. 1811): 63–64; Agnes Addison [Gilchrist], "Early American Gothic," in George H. Boas, ed., *Romanticism in America* (Baltimore: Johns Hopkins, 1940), pp. 118–36; Helen M. P. Gallagher, *Robert Mills* (New York: Columbia University Press, 1942), pp. 37, 211–13; "Dorsey's Gothic Mansion," *Port Folio* 5, no. 2 (Feb. 1811): 125–26. The articles in *Free Masons Magazine* and *Port Folio* and the two

ings were destroyed within a few years, and Latrobe himself finally summed up the effects of his experiment as deplorable, primarily because Americans—including his own pupils—were, in his judgment, incapable of understanding or constructing Gothic. "The Bank of Philadelphia has done more mischief than that of Pennsylvania has produced good. The Free Masons' Hall, which is anything but Gothic, has made me repent a thousand times that I ventured to exhibit a specimen of this architecture. My mouldings & window heads appear in horrid disguise from New York to Richmond." [32]

In the midst of this architectural innovation, Charles Brockden Brown published the first American Gothic novels. Brown came from a Philadelphia Quaker family and, although he spent some of his most productive years in New York, set his most important novels in Pennsylvania. He was again living in Philadelphia during Latrobe's Gothic years. Brown's six novels imitated the sensational best-sellers of Mrs. Radcliffe and "Monk" Lewis; but, rather than use their medieval artifices for creating terror, such as gloomy, labyrinthine castles and superstitions of diabolical influence, he chose to depict the terror of the American environment and of actual murders, Indian raids, and catastrophic epidemics. A particularly effective example of his use of American natural objects to arouse the same emotions stirred by Mrs. Radcliffe's imagined ruins occurs in *Edgar Huntly*, where a young man views the awesome scenery of the Delaware Water Gap while fleeing from marauding Indians, exhausted, in pain, lost in the forest at night, and apprehensive for the safety of his family. The authenticity of this hair-raising episode, which has fascinated numerous readers and drawn comment from most of Brown's critics, was attested to by Brown himself, who had visited the area in 1793. He asserted in an advertisement, "Those who have ranged along the foot of the Blue-Ridge from the Wind-Gap to the Water-Gap will see the exactness of the local descriptions." This statement was verified by one of the present authors, a resident of the area, who recognized the

site in the novel, traced Edgar's route on a topographic map, and identified a number of other scenes in the story, including the cave that dominates the first half, by following that route as closely as modern roads permit. [33]

Although best known for his realistic depiction of American scenes, Brown was by no means uninterested in reconstructing an imaginative medieval past. In his youth he translated Macpherson's prose version of Ossian into verse (now lost), and his friends recalled that he spent hours drawing plans for and sketching imaginary Gothic buildings. At some time he wrote two lengthy fragments, published only after his death, on the history of an English family in medieval and early modern times. "Sketches for the History of Carsol" and "Sketches for the History of the Carrils and the Ormes" have puzzled most students of Brown, who have dismissed them either as juvenilia or as attempts at utopian romance. W. T. Berthoff, however, has proposed that they were written, or at least revised, after the publication of his novels, and that they therefore represent, along with his political pamphlets and compilations of current history, an enlargement of his literary interests. If this is the case, it is possible that they were written as potential filler for Brown's magazine, *The Literary Magazine and American Register,* and as such were never intended to be carried farther, being, rather, examples of the short-lived but recognized genres of "sketch" and "Gothic fragment" used in periodicals before the classic short story form was crystallized by Edgar Allan Poe. If this is the case, these fragments must have been written in Philadelphia at the time when Latrobe was erecting one of that city's principal buildings in the Gothic style. [34]

These sketches relate the history of a dynasty roughly paralleling and, at times, intersecting that of the ruling families of England. "The Carrils and

descriptions of abbeys are listed in J. Meredith Neil, "Architectural Comment in American Magazines, 1783–1815," in *American Association of Architectural Bibliographers Papers,* ed. William B. O'Neal (Charlottesville: University Press of Virginia, 1968), 5:17–45.

[32] Latrobe to David Hare, May 20, 1813, quoted in Hamlin, *Benjamin Latrobe,* p. 248.

[33] See William Dunlap, *Life of Charles Brockden Brown,* 2 vols. (New York: J. P. Parke, 1815), 1:57–58; Donald A. Ringe, *Charles Brockden Brown* (New York: Twayne, 1966), chap. 5; R. W. Stone, "Northampton County," in *Pennsylvania Caves* (2d ed.; Harrisburg: Pennsylvania Topographic and Geological Survey, 1932). William Book and Eugene Genay of the Bethlehem Cave Club and Laura Gottshall assisted in identifying the locale of *Edgar Huntly.*

[34] See Harry R. Warfel, *Charles Brockden Brown* (Gainesville: University of Florida, 1949), p. 25; W. B. Berthoff, "Charles Brockden Brown's 'Historical Sketches': A Consideration," *American Literature* 28, no. 2 (May 1956): 147–54; Devendra P. Varma, *Gothic Flame* (New York: Russell & Russell, 1966), pp. 186–88.

the Ormes" surveys 1,600 years, beginning with Arthur, "tenth prince of Artland," a native prince in Roman Britain and father of Helena, the mother of Constantine. (Neither this Arthur nor Artland have any historical existence.) Important medieval members of the family were two sainted bishops, both named Arthur Carril, whose careers resembled that of Thomas à Becket, and Pamphila, a formidable fifteenth-century "granddaughter" of Henry IV, whose religious devotion prompted her to found an order, to build churches, to interfere in ecclesiastical politics, and, finally, to believe herself the reincarnation of the local saint for whom she was named. In the eighteenth century the family split into several branches, involving themselves in Richardsonian complications as they tried to unite and extend their lands by intermarriage with not-always-willing cousins. The central subject of both "The Carrils" and "Carsol," which deals with a branch of the same family on a Mediterranean island during the Renaissance, is the origin and transformation of religious superstitions, particularly those prompting the building of churches and convents, buildings that Brown described in exhaustive detail by imaginatively reconstructing them from their ruins.[35]

The use of history in these sketches reflects the level of historical understanding among novel-readers of Brown's time, who were familiar with a few colorful personalities and events but had no clearly defined sense of period, chronology, or history as movement in time. For them, the past was an exotic stage setting against which characters very like themselves enacted thrilling adventures. But Brown was less effective in imagining such settings than in arousing thrills by depicting landscapes he had seen and that any reader might visit. His detailed description of Pamphila's Gothic convent, for example, reflects the linear style of his fanciful architectural drawings far more than the mass of an actual building, while Edgar Huntly's view of the Delaware Water Gap conveys to readers a vivid picture of a striking geographical feature and can arouse in the tourist an additionally thrilling shock of recognition. Indeed, Brown's capacity for vivid description won the admiration of Sir Walter Scott, although Scott was repelled by Brown's explorations of abnormal psychology,

which influenced Shelley, Keats, and Peacock. "Brown had wonderful powers, as many of his descriptions show[,] but I think he was led astray by falling under the influence of bad examples, prevalent at his time. Had he written his own thoughts, he would have been, perhaps, immortal; in writing those of others, his fame was of course ephemeral."[36]

Belligerent incidents at sea and on the frontier, the embargo, and the War of 1812 aroused violently anti-British feelings in many parts of the United States between 1807 and 1815, during which period the only important Gothic building to be constructed outside New England was Saint Patrick's Church in New York City, designed by a Frenchman in accordance with long-standing Catholic tradition. However, in New England, where prosperity depended on trade with England, many resented the embargo and were dubious about the wisdom of the war. The Hartford Convention advocated secession from the Union over this issue. It is therefore interesting to note that New Englanders were apparently more receptive to Gothic architecture in this period than were other Americans. In 1809 the Congregationalist flock of William Ellery Channing commissioned Charles Bulfinch to build the Gothic Federal Street Church in Boston. This departure from the style traditional for meetinghouses caused some of Channing's neighbors to raise their eyebrows, as indeed did many of Channing's romantic religious ideas, which had already made him a rebel against Puritan orthodoxy and which would soon lead him into the Unitarian schism. For example, the Reverend William Bentley of Salem commented, with more than a hint of Yankee irony, "This is the first attempt at a Gothic Church in New England. We have had Gothic theology for many generations, and the style is not yet lost." Another remark appears, to modern eyes, to damn with faint praise, but more probably the observer merely shared the confusion then rampant about the proper subdivisions of Gothic. "It is a fine specimen of Saxon Gothic . . . admirable for its uniformity and the symetry [*sic*] of its proportions."[37] The Episcopalians of New England promptly took this church

[35] Brown, "Sketches for the History of Carsol" and "Sketches for the History of the Carrils and the Ormes," in Dunlap, *Brown,* 2:170–338.

[36] Scott quoted in Lulu R. Wiley, *Sources and Influences of the Novels of Charles Brockden Brown* (New York: Vantage, 1950), p. 271.

[37] Harold Kirker, *Architecture of Charles Bulfinch* (Cambridge, Mass.: Harvard University Press, 1969), p. 249.

150

FIG. 12. Washington Allston, *The Flight of Florimell.* Boston, 1819. Oil on canvas; H. 36″, W. 28″. (Detroit Institute of Arts, City Appropriation Purchase.)

as a model, imitating it in St. John's, Providence (1810), and in Trinity, New Haven (1814).

Channing's appreciation of the artistic side of religion was shared by the painter Washington Allston, who married Channing's sister the year the Federal Street Church was completed, and who, during his brief honeymoon residence in Boston, painted portraits of the entire Channing family, with whom he had been intimate since his schooldays. In 1811, in spite of the threat of war, Allston returned to the wider artistic opportunities of London and stayed there until 1818, maintaining his friendships with Coleridge and other romantic writers. His favorite subjects were episodes from scriptural history, particularly miracles and other events evoking supernatural terror or awe; but during and immediately after his stay in England he painted a few medieval scenes to illustrate the chivalric romances of Spenser, Shakespeare, and Ariosto. *The Flight of Florimell* (Fig. 12) depicts an incident from *The Faerie Queene;* the delicacy of

the terrified heroine, the power of her galloping horse, and the brooding mystery of the enchanted forest through which she flees from her evil pursuers evoke the fusion of pity and terror cherished by Mrs. Radcliffe's devotees. The subtle green and gold magic of this scene of knightly adventure creates a medieval world altogether different from, but not more imaginative than, West's straightforward, colorful depiction of the legendary, heroic Middle Ages described by Froissart.[38]

Allston's coterie in London included Samuel F. B. Morse and C. R. Leslie, the last of a generation of American painters who had come to England to study with West; actor-dramatist John Howard Payne; and poet Leigh Hunt (a grandnephew of Mrs. West). In 1815 they were joined by Washington Irving, who a decade before in Italy

[38] See Edgar P. Richardson, *Washington Allston* (Chicago: University of Chicago Press, 1948); Flexner, *Light of Distant Skies,* pp. 124–42, 176.

had considered studying painting with Allston and who had come to England on business as soon as the war was over. Irving approached literature, drama, and the visual arts as forms of recreation for the leisure of a merchant gentleman, an approach that had served him well as an occasional author in mercantile New York and one that would serve him well again in England when, after the failure of the family business, he turned to letters as a means of support. In *The Sketch Book* he compared his observations of traditional English culture with the conceptions of it that he had derived from books, describing in particular detail his impressions of such literary landmarks as Westminster Abbey, Shakespeare's birthplace, and the site of Falstaff's Boar's Head Tavern in Eastcheap. His form and style were themselves a part of the tradition he depicted, being derived from those of Addison, who, though respected as a great literary figure of the past in England, was "still a living author in New York."[39] But Irving was fully aware that his distance in time and space from both the tradition he practiced and the tradition he described afforded him a dimension beyond Addison's reach. Never before had English readers been given such a delightful opportunity to look at themselves through the eyes of a sympathetic foreigner, and they responded at once by taking Irving to their hearts and in the long run by accepting *The Sketch Book* as a favorite text for teaching English to speakers of other languages.

Among Irving's discoveries in England were medieval ruins and their Gothic imitations, concerning which his principal mentor was none other than Sir Walter Scott. Before Irving had come to England, Scott had enjoyed *Knickerbocker's History of New York,* which he had compared, in its humor, to Swift (whose works he had edited) and to Sterne. Perhaps he also valued it for its use of the sort of local folklore of which he himself was such an avid collector. In 1817 Irving visited Abbotsford while touring Scotland and was most hospitably entertained, although Scott was, at the time, in the midst of rebuilding his mansion and finishing *Rob Roy.* When, after Scott's death, Irving recounted this visit in a long essay, "Abbotsford," he described the house and his trip to the ruins of Melrose Abbey, and concentrated on his host's lordly generosity of character and his interest in medieval

relics and Scottish folklore. Their friendship was reinforced when Scott assisted Irving in publishing *The Sketch Book,* a fact made known to the world because Scott's sponsorship of the work encouraged a suspicion that it, like every other pseudonymous book that appeared in those years, was yet another of the numerous productions of "The Great Unknown." Irving then complimented his benefactor by using him as the model for the squire in *Bracebridge Hall,* an antiquarian lord of the manor who amiably insists on reviving falconry, village sports, and other folkways of Merry England much to the amusement of his family and tenants, who accept his eccentricities as the anachronistic crotchets of a beloved old man. In spite of this good-humored satire of his friend's foibles, Irving, already America's pioneer folklore-collector, was keenly aware of the importance of the remnants of traditional customs as an equivalent, in human behavior, to the ruins of ancient buildings. "Nothing in England exercises a more delightful spell over my imagination, than the lingerings of the holiday customs and rural games of former times. . . . They resemble those picturesque morsels of Gothic architecture, which we see crumbling in various parts of the country, partly dilapidated by the waste of ages, and partly lost in the additions and alterations of later days."[40]

While Irving profited from the example of the early Scott, transmuting folklore into poetic romance, James Fenimore Cooper earned the sobriquet "The American Scott" by modeling thrilling tales of adventure on Scott's later historical novels. These two authors met cordially in Paris in 1826, and this time it was the American's turn to offer the distressed Scott practical assistance concerning the possible recovery of royalties from his trans-Atlantic audience. Their philosophies of history diverged widely; this became particularly evident when Cooper published a series of novels set in the Middle Ages in which he directly challenged Scott's interpretation of the period. As Van Wyck Brooks explained:

His plan, like Mark Twain's in the *Connecticut Yankee,* was to exhibit American ways in the light of European

[39] Brooks, *World of Washington Irving,* pp. 161, 153–56.

[40] Irving, *Sketch Book* (New York: Collegiate Society, 1905), pp. 254–55. See also Brooks, *World of Washington Irving,* pp. 159–61; Pierre M. Irving, *Life and Letters of Washington Irving,* 4 vols. (New York: Putnam, 1863), 1:240, 380–86, 2:19–24; Irving, "Abbotsford," in *Crayon Miscellany* (New York: Collegiate Society, 1905).

history, and the three novels were pictures of late mediaeval society in Italy, Switzerland and Germany, as a
democrat saw it. Entering the feudal world of Scott,
Cooper intended no doubt to break the spell that Scott
had woven about it, and these novels defended the rule
of the people against irresponsible oligarchies who ques-
tioned the capacity of men to govern themselves. They
showed the evil of institutions that throve on the ignorance of the masses and had no proper base in the will
of the nation.

In *The Heidenmauer,* Cooper visualized this theory particularly effectively and then went beyond
it. The story takes place near a Rhenish town, in a
complex of ruins including a castle, an abbey, and
a Roman camp, with a primeval rock formation in
the midst of them. First the ruins are described by
a nineteenth-century tourist familiar with their historical context, and then they become the scene of
a sixteenth-century story whose characters, though
unacquainted with history, gather from the ancient
remains a shadowy awareness of the reality of the
past. The structures themselves contribute to the
events in the story, whether military maneuvers on
the ramparts of the castle, awesome ceremonies in
the abbey church, or trysts among the "haunted"
Roman ruins. In this setting, Cooper depicts feudal
and monastic tyranny and then goes on to demonstrate that "reforms" can be self-defeating—as when
the incensed burghers, disquieted by Lutheranism
but immediately instigated by a greedy local lord,
rise up against the oppressive monks and sack the
abbey, reducing the venerable cloister to rubble
and leaving the majestic chapel in flames.[41]

Of the many American writers of this generation
who imitated Scott in one way or another, John
Pendleton Kennedy was the most successful in applying medievalism to American materials. In his
first novel, *Swallow Barn,* a Virginia version of
*Bracebridge Hall,* he contrasted medieval survivals
in the gentry's way of life—including their system
of slavery, depicted on the eve of Nat Turner's re-
bellion as a feudal exchange of service for protection—with their children's deliberate attempts to
revive the medieval customs described in Scott's
novels—notably falconry, minstrelsy, and courtship
practices. Then, in *Rob of the Bowl,* Kennedy's re-

search unearthed materials on the history of seventeenth-century Maryland that permitted him to
describe the proprietary era in his native state as
medieval in Scott's sense. He envisioned a living
town on the ruins of the original capital at St.
Mary's, which he visited not long after the demolition of the last building, and depicted the titled
Calvert family as American exponents of feudal
social order and chivalric behavior. He described
the leaders of the one coherent Catholic culture in
the thirteen colonies both as victims of superstition
and as noble advocates of toleration for all sufferers
from persecution. The addition of a fictitious pair
of lovers who practiced falconry, minstrelsy, and
courtly love as inherited traditions completed this
picture of a medieval way of life instantly recog-
nizable to admirers of Scott, yet solidly founded on
the facts from Maryland history.[42]

Although Kennedy's intention in *Swallow Barn*
was, like that of his model, Irving, to satirize young
people who took Scott literally, many southern
readers then and later did take Scott seriously as a
cultural arbiter. Kennedy himself contributed to
this tendency, one example being a chapter entitled "Traces of the Feudal System," by which he
meant the way of life of the seventeenth-century
English gentry. The following passage from that
chapter contains the germs of the Cavalier myth
with which later Virginians assured themselves that
their peculiar culture had been created by descendants of gentlemen whose blue blood had been further refined by residence in the Old Dominion.
"[Virginia's] early population, therefore, consisted
of gentlemen of good name and condition, who
brought within her confines a solid fund of respectability and wealth. This race of men grew vigorous
in her genial atmosphere; her cloudless skies quickened and enlivened their tempers and, in two centuries, gradually matured the sober and thinking
Englishman into that spirited, imaginative being
who now inhabits the lowlands of this state." The
history of the consequent "translation of the theme
from Abbotsford into an all-pervading way of life"
in the South has been written by Rollin G. Osterweis, who concludes that, though literary magazines throughout the section published similar imitations of Scott's works, the enactment of customs

[41] Robert E. Spiller, *Fenimore Cooper: Critic of His Times*
(New York: Russell & Russell, 1963), pp. 116–18; James
Grossman, *James Fenimore Cooper* (Palo Alto: Stanford Uni-
versity Press, 1949), pp. 53–57, 124–30; Brooks, *World of
Washington Irving,* p. 336; Cooper, *The Pathfinder: The
Heidenmauer* (Boston: Brainerd, n.d.), 265–305.

[42] Kennedy, *Rob of the Bowl* (Hildesheim: Ohms, 1969),
pp. 7–19. See Henry C. Forman, *Architecture of the Old
South: The Mediaeval Style, 1585–1850* (1948; reprint ed.,
New York: Russell & Russell, 1967), pp. 107–12.

derived from them shows definite regional differences resulting from differences in ethnic composition, economy, and previous tradition. Thus Virginia concentrated on acting out the chivalric code, as in "tournaments," South Carolina emphasized romantic nationalism, Louisiana added a French quest for imperial grandeur, and the Texas frontier reduced chivalry to popular terms, particularly in religion and folklore. "Others read the *Waverley Novels* with enthusiasm but the South sought to live them." [43] The myth of chivalry, like the institution of slavery and the cultivation of cotton, became a symbol of stability for a South that had in all other respects been transformed beyond recognition between 1815 and 1860.

The translation of Scott's ideals into bricks and mortar came, as one might expect, more slowly than literary imitation of his works. Until the 1830s, Gothic architecture was most familiar to Americans as a style of Episcopal churches, very often selected by rectors of British extraction, such as John Henry Hopkins, who built the first Gothic church in the West at Pittsburgh, in 1823. Thirteen years later Hopkins, then bishop of Vermont, published the first American book on ecclesiastical Gothic, in which he advocated its use as a style of decoration tending to evoke feelings of reverence and awe. Not until five years later was the liturgical and architectural "ecclesiological revival" introduced to American Episcopalians by Bishop George W. Doane and the architect Richard Upjohn. This movement, which has been thoroughly studied by Phoebe M. Stanton, falls outside the scope of this paper because its approach to the medieval past was archaeological rather than imaginative; but it should be remarked that American ecclesiologists envisioned Gothic as a symbol of denominational identity rather than as a starting point for the moral regeneration of a rapidly industrializing nation, as did their Anglican counterparts.[44]

As least as important as the controversial and therefore thoroughly documented ecclesiological movement was the adoption of the Gothic style, at about this time, for numerous churches of various denominations with an inherited distrust of anything suggesting popery. This Gothic was still ornamental rather than structural, for pointed arches over doors and windows, plaster vaulting, and spires conveyed a satisfying sense of enhanced social status to worshipers who would have recoiled from a deep chancel or a raised stone altar as repugnant to their theological convictions. Most such churches were constructed essentially as auditoriums for the large audiences of popular preachers whose sermons conveyed spiritual truths to the imagination through emotional excitement, much more in the manner of the political oratory of the day than of eighteenth-century religious address. A conspicuous example was the Washington Square Dutch Reformed Church in New York City, built for the Reverend James Matthews by Minard Lafever. Dr. Matthews was also chancellor of New York University when Ithiel Town and Alexander J. Davis erected its Gothic campus, also on Washington Square, focused on a tremendous chapel modeled after the Great Hall of Henry VIII at Hampton Court (Fig. 13). Its immense traceried windows, filling an entire wall, and the spectacular fan vaulting, thus dramatically illuminated, created a fit setting for Dr. Matthews's eloquence, and the architecture must have been at least as overwhelming and instructive as the sermon to students who had previously worshiped in simple, if not starkly plain, town and country churches. The incongruity of adopting for religious purposes the flamboyant extravagance of a late Gothic monument exalting the secular splendor of a Renaissance monarch may strike a modern observer as ironic, but the contradiction was probably not apparent to enthusiastic Americans in a decade of unprecedented economic expansion. The panic of 1837 rudely awakened trustees, and the rest of the nation, to financial realities; the extravagant Matthews was ignominiously dismissed from his just-completed, heavily in debt campus, and the ambitious Gothic plans that Town and Davis had prepared for several western colleges were never executed.[45]

[43] Kennedy, *Swallow Barn* (New York: Harcourt, 1929), p. 57; Osterweis, *Romanticism and Nationalism*, pp. 213, 215.

[44] Agnes Addison [Gilchrist], "Early American Gothic," pp. 118–31; Hopkins, *Essay on Gothic Architecture* (Burlington, Vt.: Smith & Harrington, 1836); Stanton, *Gothic Revival and American Church Architecture* (Baltimore: Johns Hopkins, 1962), chaps. 1–3 passim.

[45] See Carol E. Hoffecker, "Church Gothic: A Case Study of Revival Architecture in Wilmington, Delaware," in *Winterthur Portfolio 8*, ed. Ian M. G. Quimby (Charlottesville: University Press of Virginia, 1973), pp. 215–31; James Early, *Romanticism in American Architecture* (New York: Barnes, 1965), chap. 4 passim; Jacob Landy, *Architecture of Minard*

FIG. 13. Ithiel Town and Alexander J. Davis, the proposed chapel, of New York University. New York City, 1837. Watercolor by Davis; H. 26″, W. 19¼″. (New-York Historical Society, A. J. Davis Collection.)

FIG. 14. Sunnyside, Tarrytown, N.Y., after 1835. (Sleepy Hollow Restorations.)

Although the "dim religious light" surrounding ecclesiastical ruins in Scott's works probably helped to condition non-Episcopalian readers to accept Gothic and its concomitant feelings of wonder and awe as suitable settings for worship, his influence on American domestic architecture was much more direct. Signs of this influence first appeared when Americans who had visited Abbotsford returned with plans to rebuild their own modest colonial country houses in its image. Irving, who had seen Scott's home under construction, imitated its spirit rather than its style, imaginatively combining seventeenth-century Dutch and Tudor elements from the traditions of the locality to create at Sunnyside "a little old fashioned stone mansion, all made up of gable ends, . . . as full of angles and corners as an old cocked hat," and in time covered with ivy from slips brought from Abbotsford (Fig. 14).[46] Cooper,

who had met Scott away from his home and who preferred the Continent to England, was content to adorn Otsego Hall with a row of battlements that looked rather incongruous on an unpretentious New York State farmhouse and furthermore proved impractical because they retained the heavy snows characteristic of Cooperstown, causing the roof to leak (Fig. 15).[47] The Baltimore merchant and art patron Robert Gilmore, returning from a pilgrimage to his ancestral Scotland, conceived the most ambitious imitation of Abbotsford. He commissioned Alexander J. Davis to design this new home, which he decided to call Glenellyn, a name beloved by admirers of *The Lady of the Lake*. With a castellated facade and tremendous pointed windows, Glenellyn was more completely Gothic than any of the others and was the first American house to carry its Gothic decoration throughout the interior (Fig. 16). Its delicate beauty was enhanced by grounds landscaped in the style of Humphry Rep-

---

*Lafever* (New York: Columbia University Press, 1970), pp. 71–74; Roger H. Newton, *Town and Davis, Architects* (New York: Columbia University Press, 1942), pp. 229–39.

[46] Irving quoted in Edward Wagenknecht, *Washington Irving: Moderation Displayed* (New York: Oxford, 1962), p. 21. See also "Sunnyside Edition," *American Collector*, vol. 16, no. 9 (Oct. 1947); Newton, *Town and Davis*, pp. 219–20.

[47] See Henry W. Boynton, *James Fenimore Cooper* (New York: Century, 1931), pp. 251–53; James F. Beard, ed., *Letters and Journals of James Fenimore Cooper*, 6 vols. (Cambridge, Mass.: Harvard University Press, Belknap Press, 1960–68), 3:9, 56–58, 152–55, 159–61, 213–25.

FIG. 15. Charles E. Thomas, *Otsego Hall*. Cooperstown, N.Y., ca. 1890. Watercolor; H. 10¼″, W. 19¼″. (New York State Historical Association.)

ton, and approached by way of a gatehouse in-geniously designed as an imitation ruin.[48]

The fashion for castellated mansions soon spread far beyond wealthy tourists' imitations of Abbots-ford. It was often suggested in the 1830s that the scenery of the lower Hudson needed only castles, with their historic associations, to be equal or su-perior to that of the Rhine, and New York mag-nates, taking this suggestion seriously, undertook to provide the castles. A number of these were de-signed by Town and Davis, the most conspicuous being Lyndhurst, built in 1838 by merchant gentle-man William Paulding, father of a mayor of New York and brother-in-law of Washington Irving. The original house, patterned after Lowther Cas-tle, which Town had recently seen in England, was in the pointed Gothic style rather than the perpen-dicular Gothic of Glenellyn. It boasted a vaulted central hall, several additional gables with carved bargeboards and ogival windows, and a piazza re-calling an arcaded cloister. In the 1860s Davis com-

pletely remodeled this building, doubling its size, for a new owner, so that the Lyndhurst now open to the public by the National Trust for Historic Preservation, while without question the most sig-nificant museum of American Gothic, displays, for the most part, aspects of the revival beyond the scope of this paper.[49] Not all observers admired these efforts to increase the picturesque character of the Hudson Valley, as can be seen in this anony-mous review of 1846.

It was quite pardonable in Horace Walpole and Sir Walter Scott to build gingerbread houses in imitation of robber barons and Bluebeard chieftains; they were poets and had written Gothic romances; they would fill their houses with rusty old armour, lances, drinking horns and mouldy tapestry, and they were surrounded by the memorials of the times they were idly trying to revive. But there can be nothing more grotesque, more absurd, or more affected, than for a quiet gentleman, who has made his fortune in the peaceful occupation of selling calicoes, and who knows no more of the middle ages than they do of him, to erect for his family resi-

[48] See Newton, *Town and Davis*, pp. 214–17; Wayne An-drews, "America's Gothic Hour," *Town and Country* 101, no. 4302 (Nov. 1947): 144 ff.

[49] Newton, *Town and Davis*, pp. 218–22; Helen Duprey Bullock, "Paulding's Manor or Paulding's Folly: The 1838–64 Era of Lyndhurst," in *Lyndhurst* (Washington, D.C.: Na-tional Trust, 1973), pp. 28–31.

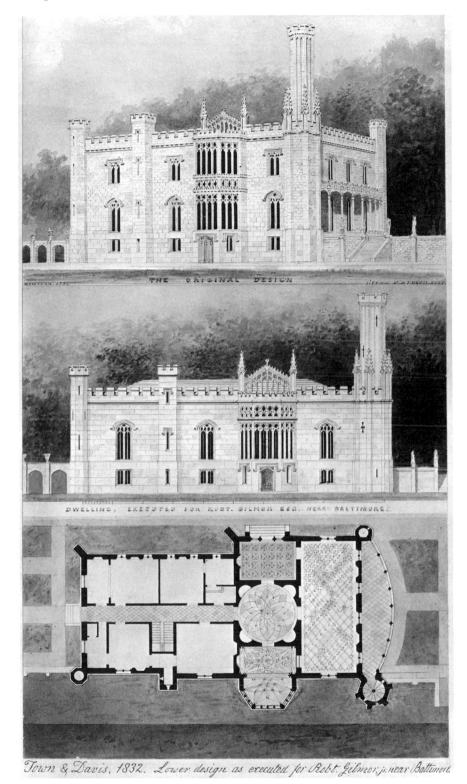

FIG 16. Alexander J. Davis, elevation and plan of Glenellyn. Baltimore, 1832. Watercolor; H. 15 5/16″, W. 8½″. (Metropolitan Museum of Art, Harris Brisbane Dick Fund.)

Fig 17. Andrew Jackson Downing, *Interior in a Simple Gothic Style*. From Downing, *The Architecture of Country Houses* (New York: Appleton, 1850), p. 383, fig. 178.

dence a gimcrack of a Gothic castle . . . as though he anticipated an attack upon his roost from some Front de Boeuf in the neighborhood.[50]

Most of these houses, like most Gothic homes in England, were obstinately eighteenth century in their internal arrangements and contemporary in their furnishings, for the very good reason that little was known about medieval interiors (Fig. 17). The furniture that survived from the Middle Ages was, as a rule, for public purposes, such as thrones and church fittings. Chippendale and his English successors adapted decorative motifs from these and from Gothic architecture, although many such motifs were more properly ecclesiastical than domestic. The revival of medieval craftsmanship by John Ruskin and William Morris was still in the future, as was the scholarly demonstration that medieval halls and their furnishings were designed for a way of life in which privacy and domestic comfort, taken for granted in the nineteenth century, were unheard-of. Nevertheless, Davis was outstanding for his meticulous care in decorating the interiors of his Gothic houses in keeping with their exteriors, as he did at Lyndhurst, both times with the assistance of the talented English-born cabinetmaker, Richard Byrnes. Davis's furniture designs were based on Chippendale's Gothic, but where

Chippendale in effect applied elaborate decoration to the surface of his pieces, Davis, as in his buildings, integrated ornamentation with structure, as in the chairs from the original furnishings of Lyndhurst (Fig. 18). They present an imaginative combination of the picturesque and the practical, particularly in the suggestive yet substantial rose window backs.[51]

This custom-made furniture, like the houses for which it was designed, was sufficiently elaborate and expensive to serve the wealthy as a status symbol. But in the same years Andrew Jackson Downing, the first widely read American writer on architecture, called attention to Gothic as a style far more commodious, convenient, and comfortable than Greek revival for tasteful homes for the common man. Downing, who visited England only after the publication of his most influential works, derived his idea of Gothic from English books, particularly John Britton's series describing historic buildings and Augustus Charles Pugin's and E. J. Willson's *Specimens of Gothic Architecture*. In his books, notably *The Architecture of Country Houses*, Downing presented Gothic designs for all types of homes, from simple laborers' cottages to elaborate villas requiring numerous servants, along with estimates of probable expenses and directions sufficient for the guidance of small-town builders. He called his cottages for working people, built of wood with pointed gables, casement windows, and carved bargeboards, "bracketed," reserving the label Gothic for a more elaborate stone dwelling with an asymmetrical plan, an oriel window, clustered chimneys, and an arcaded entrance (Fig. 19). This design Downing recommended for a gardener's cottage or gatekeeper's lodge, "where sufficient architectural style to harmonize with the general air of the estate is permissible," but he suggested that it could be made more in keeping with an ordinary neighborhood by squaring off the pointed attic windows and simplifying the bay window. The plan showed a living room and a kitchen on the first floor, three bedrooms on the second, and an attic and a cellar; Downing estimated that it could be built with common stone trim for under

[50] *New York Mirror,* Oct. 17, 1846, quoted in Talbot Hamlin, *Greek Revival in America* (1944; reprint ed., New York: Dover, 1964), p. 325.

[51] See Siegfried Giedion, *Mechanization Takes Command* (New York: Oxford, 1948), pp. 258–72, 299–304, 318–28, 344–63, 385–88; Newton, *Town and Davis,* pp. 113–18, 238; Metropolitan Museum of Art, *Nineteenth-Century America: Furniture and Other Decorative Arts* (New York: New York Graphic Society, 1970), no. 97.

Fig. 18. Alexander J. Davis, wheelback chairs and table with round top designed for Lyndhurst, Tarrytown, N.Y. Probably New York, 1838–41. Oak; chairs, H. 37″, W. 18½″, D. 17½″; table, H. 29¼″, Diam. 34½″. (National Trust for Historic Preservation.)

Fig. 19. Andrew Jackson Downing, *Small Gothic Cottage*. From Downing, *The Architecture of Country Houses* (New York: Appleton, 1850), p. 93, fig. 18.

FIG. 20. John Pickering house, Salem, Mass., 1660; remodeled 1841. (Essex Institute.)

$1,000, as compared with $400 for his simplest two-room laborer's cottage and about $1,200 for his least-expensive farmhouse.[52]

Downing houses soon began to spring up along the Hudson and in many frontier towns whose residents, having worked their way off the bare edge of subsistence, were beginning to be concerned with the appearance of their homes. Indeed, the fashion became so insistent that existing colonial houses, particularly in New England, received a face-lifting of Gothic trim. The Wedding-Cake House in Kennebunkport, Maine, is the best-known example, but another of particular interest is the John Pickering house in Salem, Massachusetts, a multigabled seventeenth-century example

of medieval survival made consistent with nineteenth-century medievalism by the superimposition of jigsawn bargeboards (Fig. 20). Let us hope that the residents of these houses carried out Downing's theory that a Gothic home enhanced the quality of life within it by evoking wholesome "associations" with cultural traditions of the past.

The sight of an old English villa will call up in the mind of one familiar with the history of architecture, the times of the Tudors, or of "merry England" in the days of Elizabeth. The mingled quaintness, beauty and picturesqueness of the exterior, no less than the oaken wainscot, curiously carved furniture and fixtures of the interior of such a dwelling, when harmoniously complete, seem to transport one back to a past age, the domestic habits, the hearty hospitality, the joyous sports, and the romance and chivalry of which, invest it, in the dim retrospect, with a kind of golden glow, in which the shadowy lines of poetry and reality seem strangely interwoven and blended.[53]

[52] See Downing, *The Architecture of Country Houses* (1850; reprint ed., New York: Dover, 1968), pp. 92–95, 73–78, 145–50; George B. Tatum, "Andrew Jackson Downing: Arbiter of American Taste" (Ph.D. diss., Princeton University, 1949), chap. 4; Edythe N. July, "Andrew Jackson Downing: A Guide to American Architectural Taste" (M.A. thesis, New York University, 1945), pp. 91–97.

[53] Downing, *Cottage Residences* (New York: Putnam, 1842), p. 33, quoted in Tatum, "Downing," p. 151. See also Alan

FIG. 21. Thomas Cole, *The Return*. Catskill, N.Y., 1837. Oil on canvas; H. 39¾", W. 63". (Corcoran Gallery of Art.)

In view of the concentration of Gothic buildings in the Hudson Valley, it is noteworthy that the Hudson River school of painters showed little interest in depicting either these homes or medieval imaginative fantasics, most of them being content to paint the landscape of the Catskills as they saw it. The principal exception was a set of two paintings by Thomas Cole: *The Departure,* in which a knight leaves his castle and goes off to war in the spring, and *The Return* (Fig. 21), in which his body is brought home to his ancestral church in the autumn. Though Cole, in these paintings, depicts the context of one of West's battles and appears to share Allston's awareness of the mystery of the forest, his use of scale conveys an impression of the

Middle Ages as a period in which the greatest and proudest individuals were so small as to be barely visible against the towering architectural structures of their civilization and the still more immense remnants of geological antiquity. So Cole's English heritage and American experience fused to create a conception of the Middle Ages, not as a setting for colorful events or as a source of picturesque ornament, but as one of many eras in the history of mankind, taking place in an untamed wilderness that suggests, beyond history, the dawning awareness of geologic time.[54]

Thus Cole's paintings are the ultimate example of American assimilation of the dominant image of the early English Gothic revival—crumbling ruins enveloped in an aura of heroic romance and recalling a sense of the reality of the medieval past. Americans had learned to visualize this image, not directly—by seeing ruins, relics, and historic sites—as had the English, but imaginatively—by reading

Gowans, *Images of American Living* (Philadelphia: Lippincott, 1964), pp. 310–12; Clay Lancaster, "Three Gothic Revival Homes in Lexington, Kentucky," *Journal of the Society of Architectural Historians* 16 (Jan.–June 1947): 13–21; I. T. Frary, *Early Homes of Ohio* (1936; reprint ed., New York: Dover, 1970), pp. 135–36; Theodore Bolton, "Gothic Revival Architecture in America: The History of a Style," *American Collector* 17, nos. 3, 4, 6 (Apr.–June 1948): 6–9, 15–18, 16–19. We are indebted to Frank O. Spinney of the New York State Historical Association for referring us to the example of the Pickering house.

[54] See Louis L. Noble, *Life and Works of Thomas Cole* (Cambridge, Mass.: Harvard University Press, 1964), chap. 24 passim; Donald A. Ringe, *Pictorial Mode* (Lexington: University of Kentucky, 1971), pp. 148–54, 205–6.

novels, poems, and travelers' descriptions. Therefore, although they read the same books, English readers learned from them what to look for in their historic surroundings, while American readers learned how to make their surroundings look historic. In this process, the most important task for the English was to develop a renewed awareness of the medieval tradition revealed in, what were for them, everyday objects, while for the Americans it was to develop a context of feeling and aesthetic response within which references to such a tradition, and occasional views of such objects, would be meaningful.

Awareness of the Middle Ages grew gradually in England, in accordance with the massive tradition of cultural continuity that Tennyson referred to in the following lines:

> A land of settled government,
> A land of just and old renown,
> Where Freedom slowly broadens down
> From precedent to precedent.[55]

So sixteenth-century reformers claimed to restore the religious customs of the Anglo-Saxons; seventeenth-century revolutionaries demanded the liberties confirmed by Magna Charta; and antiquarian gentlemen called attention to ruined castles and monasteries, monuments preserving the communal memory of the medieval past, as part of the national sense of identity. Addison's defense of this national character in literature against neoclassical contempt for the "barbarous Dark Ages" was reinforced by the extension of the neoclassical ideal of "nature" to include the ordered wildness of the English garden; Walpole carried this fashion for Gothic ornament to its logical conclusion at Strawberry Hill and then added a dimension to it by using descriptions of this building to enhance the excitement of a thrilling story. Historical novelists helped general readers to visualize the little history that they did know; historians wrote to inform the popular as much as the scholarly audience; and literary researchers collected great quantities of ballads and other survivals of medieval folk culture, particularly from the Celtic areas of Wales and Scotland. This literary antiquarianism, the formulation of the aesthetic of the picturesque, the emergence of tourism, and the publication of descriptions of historic sites created a rich and complex context for response to allusions to the medieval heritage, whose possibilities were fully developed by the genius of Sir Walter Scott. The unprecedented popularity of his works not only brought him wealth and renown but initiated a new era in English literature and made an indelible impression on the study, writing, and reading of medieval history.

Eighteenth-century Americans saw only shadows of this development, and even these were interrupted by war and distorted by hostility toward the mother country. The colonial impulse toward Chippendale Gothic was nipped in the bud, and Benjamin West's medieval paintings did not travel to America, despite Barlow's panegyric. Latrobe, the first of a number of immigrants who built Gothic buildings from memory, set off a chain reaction of buildings in Philadelphia that perhaps influenced Brown's pseudohistorical sketches. Then American artists visited Europe and described medieval remains; Allston illustrated poetic romance, Irving observed sites and folkways associated with literary tradition, and Cooper used the tourist's experience of ruins to illustrate conflicting issues of the past. Finally, a whole generation of Americans began imitating Scott; in literature (Kennedy), in life (Southern romantic nationalists), and in architecture (the imitators of Abbotsford, the Hudson Valley castle-builders, and Downing's converts to Gothic for the common man).

But the question remains as to why Americans should have gone to this trouble to re-create the medieval tradition, particularly when some of their own revolutionaries, such as Barlow, assured them that the Middle Ages was a period of tyranny and superstition from which America was blessedly free? It must be emphasized that all of these artists and writers worked in many styles, and that their Gothic works were comparatively few. Nevertheless, these experiments, in the face of all the difficulties attending them, suggest that medievalism offered creative opportunities that could be realized in no other way. Some, like Brown and West, found a framework for imaginative insight into historic processes and events or, like Allston and Cole, a starting point for visualizing their own feelings and evoking terror, superstitious awe, or sublime wonder. Others, like Irving and Downing, saw in medievalism an extension of memory, from individual experience to the cumulative traditions of the folk and the history of a family, a locality, or

---

[55] Tennyson, "You Ask Me, Why, Though Ill at Ease," ll. 9–12.

a cultural group. Still others found it helpful in defining their own identity, as when Cooper compared his understanding of liberty with the liberties of medieval communities. All of these functions of medievalism had developed in England and appeared in America partly because Americans of British descent could no more cut themselves off from the continuity of their cultural heritage than they could cease speaking the English language; but it soon became evident that the medieval contribution to American culture was positive in its own right. American clients commissioned Gothic buildings partly because they were fashionable abroad, but partly also because they considered it appropriate to adapt the styles found most effective in the past, under whatever system of government, to the most advanced and enlightened purposes of a society that accepted its heritage from European culture as the basis of a unique responsibility to set an example for future generations.

In some respects any kind of history would have served this purpose, and many kinds were tried, as revival succeeded revival. But why the Middle Ages? In Europe, the medieval revival was part of the wide-ranging romantic reaction against the tyranny imposed by excesses of neoclassical order, benevolent despotism, and revolution in the name of Reason; but in America formal thought, knowledge of history, and social cleavages were not sufficiently developed to permit such clearly defined

polarities. Furthermore, to Europeans the idea of the Middle Ages triggered much deliberate fantasy, created as an escape from unbearable everyday realities and consoling precisely in its remoteness from actual experience; this fantasy also appealed to Americans, whose everyday lives, although different, could be quite as unbearable, and who found consolation less in surrendering to the enchantment of a fairy tale past than in envisioning its achievements as stepping-stones toward the accomplishments of a fabulous future. This imagined Middle Ages was—and is—exceedingly durable, persisting in popular thought and feeling, and even shaping the conceptions of scholars other than those strictly confined to medieval studies, long after accurate knowledge of the period has become available. It has done great damage when its nature and proper function have been forgotten or have been confused with those of historical truth; but in the early nineteenth century it came to serve the peoples of England, Western Europe, and America as a vehicle for expressing a sense of continuity from the past, through the present, to the future. Therefore when Americans learned to use the Gothic imagery of ruins, romance, and historical reality for their own creative purposes, they not only added a new dimension to their own sense of identity but made a distinctive contribution to an important imaginative movement in the international heritage of Western culture.

# A Monument of Trade

## A. T. Stewart and the Rise of the Millionaire's Mansion in New York

*Jay E. Cantor*

IN HIS TRIBUTE to New York City, published in 1869, Junius Henri Browne enthusiastically described the motivations of the city's most eminent merchant, Alexander Turney Stewart (1803–76), a Scotch-Irish immigrant to America who turned a small dry goods business into a multimillion-dollar operation in the years immediately prior to and during the Civil War (Fig. 1). According to Browne, "More than any one else in America probably Alexander T. Stewart is the embodiment of business. He is emphatically a man of money—thinks money; makes money; lives money. Money is the aim and end of his existence, and now, at sixty-five, he seems as anxious to increase his immense wealth as he was when he sought his fortune in this country, forty years ago." Stewart was well known to his contemporaries through his commercial activities. Arriving in New York City in the 1820s, he secured a teaching position but soon abandoned education for commerce. He invested a small inheritance in laces bought on a visit to Ireland and, with the profits earned from the disposition of these goods, established himself as a merchant. He occupied a series of shops along the west side of lower Broadway, changing locations to meet the demands of his rapidly increasing business. During the depression of 1837, he bought heavily at low prices, paying cash and selling at a small markup. Realizing that volume was the key to high profits, Stewart lured customers with a barrage of handbills advertising his stock and low prices for quality goods. At a time when standard pricing was rare, Stewart introduced a one-price system for all customers and insisted on honest representation of his merchandise. Featuring a wide variety of goods, he arranged his stock according to a departmental system that became the prototype for the modern department store. Endowed with an excellent sense of timing, he anticipated growth trends and enlarged his business continually to meet the demands of a swelling market.[1]

By 1846 he was ready for the most dramatic and innovative move of his career—one that defied the conventions of successful merchandising in New York City and set a precedent for a new type of dry goods establishment. In that year Stewart opened a new marble emporium, quickly dubbed the "marble palace," on the east side of Broadway in the block between Reade and Chambers streets (Fig. 2). Although the east side of the street, called the shilling side, had traditionally been considered less

[1] Junius Henri Browne, *The Great Metropolis* (Hartford: American Publishing Co., 1869), p. 289. For contemporary descriptions of Stewart, see Matthew Hale Smith, *Sunshine and Shadow in New York* (Hartford: J. B. Burr & Co., 1868); Edward Crapsey, "A Monument of Trade," *Galaxy* 9, no. 1 (Jan. 1870): 94–101; "Alexander T. Stewart," *Harper's Monthly* 34, no. 202 (Mar. 1867): 522–24; *Frank Leslie's Popular Monthly* 1, no. 6 (June 1876). For obituaries, see *Frank Leslie's Illustrated* 42, no. 1073 (Apr. 22, 1876), no. 1074 (Apr. 29, 1876); James Grant Wilson, "Alexander T. Stewart," *Harper's Weekly* 20, no. 1009 (Apr. 29, 1876); *New York Times*, Apr. 11–14, 1876; *Evening Post* (New York), Apr. 10, 11, 13, 1876; *Sun* (New York), Apr. 11–14, 1876. Biographical descriptions and accounts, and legal documents relating to Stewart's estate and claims against it, were published in *The Posthumous Relatives of the Late Alex. T. Stewart, Proceedings before the Surrogate, Extracts from Newspapers, etc.* (New York: George F. Nesbitt, 1876). Later biographies include *Appleton's Cyclopedia of American Biography* (1888), s.v. "Stewart, Alexander T."; *Dictionary of American Biography*, s.v. "Stewart, Alexander T."; Harry E. Resseguie, "The Decline and Fall of the Commercial Style of A. T. Stewart," *Business History Review* 36, no. 3 (Autumn 1962): 255–86; Resseguie, "A. T. Stewart's Marble Palace—The Cradle of the Department Store," *New-York Historical Society Quarterly* 48, no. 2 (Apr. 1964): 131–62; Andy Logan, "That Was New York, Double Darkness and Worst of All," *New Yorker* 34, no. 1 (Feb. 22, 1958): 81–113.

FIG 1. Alexander Turney Stewart as a young man, ca. 1840. Engraving after a daguerreotype. (Picture Collection, New York Public Library.)

desirable for shopping due to its exposure to the afternoon heat and light, Stewart turned this location into an asset: the gleaming facade of his shop, with its daring street-level plate glass windows, stood out from the narrow, dark facades of his competitors' shops. The effect of the marble palace was well described by one observer who wrote:

A few years ago, when a man returned from Europe, his eye being full of the lofty buildings of the Continent, our cities seemed insignificant and mean. His first impulse was to sit upon the low roofs and dangle his feet over the street. He felt that the city had no character, but he could not see what was wanting. But the moment Stewart's fine building was erected, the difficulty appeared. That tyrannized over the rest of the street—that was a key-note, a model. There had been other high buildings, but none so stately and simple. And even now there is, in its way, no finer street effect than the view of Stewart's buildings seen on a clear, blue, brilliant day, from a point as low in Broadway as the sidewalk in front of Trinity Church. It rises out of the sea of green foliage in the Park [City Hall Park], a white marble cliff, sharply drawn against the sky [Fig. 3].[2]

Stewart's store was not only considerably larger than typical establishments; it also spread shopping through several floors. The range of his stock was staggering, and customers used to trudging from shop to shop were delighted to find that at Stewart's they could buy anything from a $4,000 shawl to a paper of pins.

The store was furnished with counters and shelves of curly maple and mahogany and was illuminated by "beautiful chandeliers manufactured by Messrs. Cornelius and Son, of Philadelphia." Additional light was provided by a domical skylight capping a dramatic well that rose the full four-story height of the interior above the basement. Fresco decorations executed by Signor Bragaldi and ornamental capitals symbolizing Commerce and Plenty carved by Ottavian Gori provided the principal embellishments. The store quickly became a showcase and New York's most fashionable bazaar. Articles appearing in newspapers and in magazines such as *Godey's Lady's Book* and *Harper's Monthly* spread its fame across the continent. In New York it was so familiar a landmark that it was unnecessary for Stewart to put a sign on the exterior identifying the structure. Standing

[2] George William Curtis, "Easy Chair Chats," *Harper's Monthly* 9, no. 50 (July 1854): 261. See also Ellen Kramer, "Contemporary Descriptions of New York City and Its Public Architecture ca. 1850," *Journal of the Society of Architectural Historians* 27, no. 4 (Dec. 1968): 270–71.

FIG. 2. J. Trench and Company, proposed Chambers Street elevation of A. T. Stewart's marble store. New York City, 1845. Watercolor. (New-York Historical Society.)

FIG. 3. A. T. Stewart's marble store, New York City, 1846. Photograph, ca. 1870–80, showing addition of a fifth story. (Museum of the City of New York.)

near the heart of the metropolis and just behind City Hall, in the vicinity of such prominent features as the Astor house and Barnum's museum, it could not fail to attract attention. Stewart enlarged the marble palace to occupy the full Broadway frontage in 1850–51, and subsequent additions carried it 225 feet into the side streets. Behind the mammoth facade of the marble palace, A. T. Stewart revolutionized retailing and made it big business.[3]

Stewart's business empire, of which he was the sole proprietor, grew to include mills and factories that supplied his retail and wholesale operations with necessities and luxury items. Having made his first buying trip to France in 1839, he had, by 1845, established a purchasing office in Paris. Eventually, offices and agents of A. T. Stewart and Company were established in Manchester, Lyons, Berlin, Chemnitz, Glasgow, Belfast, and Nottingham. He further extended his operations to include European mills, and by 1872 it was stated that one-tenth of all goods imported through the port of New York were destined for the dry goods empire of this calico aristocrat. By that time Stewart had constructed a new cast iron store further uptown, on Broadway between Ninth and Tenth streets. He had moved his retail business into this building and had retained the marble palace for his wholesale operations. Although he maintained a single retail store, he established sales offices in Boston, Philadelphia, and Chicago. The annual volume of the business rose to $40–50 million. When assessments for taxes were made during the Civil War, Stewart's personal income of $1,843,631 was the highest in the country. This was over $1 million more than the personal income of William B. Astor and $1.2 million more than that of Cornelius Vanderbilt, his nearest rivals. It is thus not surprising that Stewart's mercantile reputation was widespread or that Junius Browne chose to describe him

in such extravagant terms. But Browne's comments are of interest also because he associated Stewart with three architectural achievements: the marble palace, the cast iron store, and "his private residence, or what is designed to be such, in Fifth Avenue, corner of Thirty-fourth street." Browne had some misgivings about the last-named structure: "It is a huge white marble pile; has been four or five years in process of erection, and has already cost $2,000,000. It is very elaborate and pretentious, but exceedingly dismal, reminding one of a vast tomb. Stewart's financial ability is extraordinary, but his architectural taste cannot be commended." [4]

Junius Browne was not alone among mid-nineteenth-century writers in commenting on A. T. Stewart's house. For most contemporaries, the house occupied as significant a place in the development of palatial domestic architecture as Stewart's stores had in the evolution of commercial buildings. Yet modern historians, while according the marble palace and the cast iron store their due consideration in the history of New York's commercial architecture, have ignored Stewart's house or misunderstood the relationship between it and the mansions that followed. Indeed, there has been little detailed study of the development of the millionaire's mansion. Few modern biographies of the architects of these American palaces have been published, and the phenomenon has been left to social commentators, who have generally viewed these houses as an expression of the attitudes of the nouveaux riches. The first entry in the competition of notable mansions is usually considered to be William K. Vanderbilt's house (1879–81; Fig. 4). The Vanderbilt house was stylistically audacious in America at the time of its construction. Designed as a Francois I chateau by Richard Morris Hunt, the reigning academic master, the Vanderbilt house surpassed all earlier urban mansions in richness of materials and exterior picturesqueness. The construction of this mansion was associated with Alma Vanderbilt's campaign to win social recognition

---

[3] Resseguie, "Marble Palace," pp. 132–35, suggests that Stewart raised the status of the retailer from monger to merchant; Mary Ann Smith, "John Snook and the Design for A. T. Stewart's Store," *New-York Historical Society Quarterly* 58, no. 1 (Jan. 1974): 18–33, identifies Trench and Snook as the designers of the store and surveys the stages of construction. See also "A Sketch of the Greatest Business Man of America—How He Rules His Army of Clerks and Subordinates—The Grand System of His Business," *Cincinnati Gazette*, Sept. 16, 1875 (column headed New York, Sept. 15, 1875), reprinted in *Posthumous Relatives of the Late Alex. T. Stewart*, pp. 160–66.

[4] Testimony of Albert T. Cooper before the Senate Committee on Investigation and Retrenchment, *New York Times*, Jan. 7, 1872, quoted in Resseguie, "Decline and Fall," p. 256, n.3; Browne, *Metropolis*, p. 292. Wayne Andrews, *The Vanderbilt Legend, The Story of the Vanderbilt Family, 1794–1940* (New York: Harcourt, Brace & Co., 1941), p. 104, points out that Civil War tax figures are not necessarily an accurate index of wealth, as they were based on income and not on value of property.

FIG. 4. Richard Morris Hunt, Fifth Avenue front of William K. Vanderbilt house. New York City, 1879–81. From John V. Van Pelt, *A Monograph of the William K. Vanderbilt House,* ed. Eugene Clute (New York: by the author, 1925), pl. 21.

from Mrs. Astor. Mrs. Astor lived in a brownstone house whose exterior plainness was a discreet symbol of a tradition of cultured leisure and long-standing wealth. This dovetailing of architectural elaboration and social aspiration served to isolate the Vanderbilt mansion and its progeny from earlier, clearly established tendencies. Moreover, since these palatial houses appeared at the very moment when the opening chapters in the history of modern architecture were being written, they have seemed in retrospect all the more disconnected from a rational and evolving tradition.[5]

There is considerable documentation to affiliate these mansions and the uncultured values of the nouveaux riches. By the mid-nineteenth century, there were a number of reports on the presence of the parvenu and the demise of the old order. Abram C. Dayton, in *Last Days of Knickerbocker Life in New York,* remarked, "Thousands of the occupants of brownstone mansions which grace our leading avenues are, as is well known, not 'to the manor born.' They came from the North, South, East and West when the spirit of speculation settled upon the heretofore sleeping Dutch city." And Junius Browne amplified the image of contrast between old and new in his paean to New York City. He suggested that the main role of the new rich was to "annoy and worry the Knickerbockers, who have less money and are more stupid." He found the Knickerbockers, with their inherited wealth and conservative attitudes, "the narrowest and dullest people on the Island, [who] . . . have done much to induce the belief that stupidity and gentility are synonymous terms." On the other hand, he thought the parvenus boring. "They are exceedingly *prononcé, bizarre,* and generally manage to render themselves very absurd. . . . They outdress and outshine the old families, the cultivatedly comfortable, the inheritors of fortunes, and everybody else, in whatever money can purchase and bad taste can suggest." Edith Wharton implemented this image with an account of the architectural transgressions of the nouveaux riches. In *The Custom of the Country* she wrote:

As Ralph pushed the bolts behind him, and passed into the hall, with its dark mahogany doors and the quiet

"Dutch interior" effect of its black and white marble paving, he said to himself that what Popple called society was really just like the houses it lived in; a muddle of misapplied ornament over a thin steel shell of utility. The steel shell was built up in Wall Street, the social trimmings were hastily added in Fifth Avenue; and the union between them was as monstrous and factitious, as unlike the gradual homogeneous growth which flowers into what other countries know as society, as that between the Blois gargoyles on Peter Van Degen's roof and the skeleton walls supporting them.

To this she added a portentous image of the ultimate extinction of the "native" New Yorker at the hands of the "invaders."

Ralph sometimes called his mother and grandfather the Aborigines, and likened them to those vanishing denizens of the American continent doomed to rapid extinction with the advance of the invading race. He was fond of describing Washington Square as the "Reservation," and of prophesying that before long its inhabitants would be exhibited at technological shows, pathetically engaged in the exercise of their primitive industries.[6]

Such accounts, while accurate for a portion of the population, are too broadly drawn to account for the complexities of a highly dynamic situation. They tend to blur significant distinctions between the attitudes of those whose success was a genuine chronicle of rags to riches and the attitudes of those whose rise to prominence occurred over several generations. The subtle transition from rich to very rich and the effects of long-standing wealth on succeeding generations represent other uncounted variations. William K. Vanderbilt, for instance, as the grandson of the founder of the family fortune, was a member of the third generation of wealth. Moreover, his architect was one of the most sensitive and best-trained designers in America. Yet the house that resulted from their collaboration (and Mrs. Vanderbilt's) was different from typical homes of the wealthy that preceded it. The question is whether such differences mark changes in society or whether they merely represent changes in style to accommodate different life-styles and to submit to the tyranny of fashion.

The history of American domestic architecture throughout both the eighteenth and the early nineteenth centuries is speckled with stories of men who

[5] Alan Burnham, "The New York Architecture of Richard Morris Hunt," *Journal of the Society of Architectural Historians* 11, no. 2 (May 1952): 11; Wayne Andrews, *Architecture, Ambition and Americans* (London: Macmillan, 1964), pp. 178–79.

[6] Dayton, *Last Days of Knickerbocker Life in New York* (New York: George W. Harlan, 1889), pp. 9–10; Browne, *Metropolis,* pp. 32–35; Wharton, *The Custom of the Country* (New York: Charles Scribner, 1913, Scribner Library Paperback, n.d.), pp. 73–74.

accumulated great fortunes, often through dubious means, and built establishments that for their time were as extravagant and palatial as those constructed in the second half of the nineteenth century. All lamentations over the passing of the great families of the revolutionary era aside, the nineteenth century too was marked by an extraordinary accumulation of wealth in the hands of a relatively small percentage of the population, and it appears that this wealthy class continued to consolidate its holdings and to amass an increasingly greater portion of the aggregate wealth throughout the period. It is important to note that recent studies in economic history suggest that the differences between the early and the late nineteenth century are probably not as directly the product of the actual accumulation and distribution of wealth as has previously been thought. An important distinction in the style and ostentation of early and late nineteenth-century mansions might lie in the fact that the early ones were often rural seats rather than urban mansions. Furthermore, the great mansions constructed in the 1880s and 1890s, although urban in location, were often based on rural models and therefore did not conform to the rigid geometry of city streets. Their broken profiles and picturesque roof-massings seemed out of place when not in a landscape setting. The break with street propriety was so blatant that houses of radically different scale and material were often juxtaposed. An arrogant naïveté is assumed to be the cause of these aberrations, and, to underscore the idea that these mansions were but whims of their builders, it is pointed out that most stood for only a generation. It is profitable, however, to look at these facts from a different angle. Rather than as a vagary in a confused and competitive society, the emergence of these millionaires' mansions should be considered as a comprehensible aspect of American urban history. And it was precisely in the context of the emerging urban landscape and its institutions that one of the richest men in America, A. T. Stewart, conceived of and built his mansion.

The Stewart mansion, built in 1864–69, can be seen as a bridge between early nineteenth-century mansions and those built later in the century. In scale and expense, the Stewart house was closer to those mansions that followed it than to those that preceded it, although it was cut off from its later imitators by a hiatus in building activity during the depression of the 1870s. When construction of great houses resumed at the end of the decade,

there had emerged a new style or styles considered unrelated to that of houses like the Stewart mansion. Moreover, the Stewart house was in many ways the logical conclusion of the stylistic innovations preceding it. It therefore occupied a pivotal position, and a study of the house allows us to identify some of the emotional and symbolic roots of the later millionaire's mansion in America. Indeed, a close inspection of A. T. Stewart's career, his house, and his times reveals a logic underlying the seemingly irrational appearance of millionaires' mansions and also suggests that men like Stewart must be understood as individuals—complex and paradoxical—rather than as broadly drawn caricatures.[7]

It is important to look at the context in which Stewart operated, the milieu in which his ideas grew and assumed concrete manifestation. The New York in which Stewart made his fortune was a city of commerce. With good port facilities and a direct inland link to the Great Lakes through the Erie Canal, completed in 1825, it was becoming the commercial and financial center of the nation, just as Stewart was becoming its major merchant. James Fenimore Cooper, comparing New York to London, called the city the emporium of the New World and remarked: "New York is essentially national in interests, position, and pursuits. No one thinks of the place as belonging to a particular State, but to the United States." Yet when Charles Dickens saw the city in 1842, he described it as "low, dull, straggling and ill built." Although 4 percent of the population in 1845 commanded more than 60 percent of the city's noncorporate wealth, there was relatively little urban manifestation of this concentration of wealth or evidence of municipal improvements. Commercial expansion involved the reutilization of houses as shops and hotels, as well as the creation of new commercial structures. The pressure on downtown real estate proved tremendous, even though much of the island of Manhattan remained unoccupied. When Philip Hone, an auctioneer and former mayor of New York, moved, in 1822, to 235 Broadway, opposite City Hall Park, he paid $25,000 for a house

---

[7] For a survey of important houses of the second half of the nineteenth century, see Jacob Landy, "The Domestic Architecture of the 'Robber Barons' in New York City," *Marsyas* (1950), pp. 62–86; Edward Pessen, "The Egalitarian Myth and the American Social Reality: Wealth, Mobility, and Equality in the 'Era of the Common Man,'" *American Historical Review* 76, no. 4 (Oct. 1971): 989–1034.

FIG. 5. Philip Hone house, New York City, ca. 1825. From an engraving in Theodore S. Fay, *Views of New-York and Its Environs* (New York: Peabody & Co., 1831), pl. 4. In 1836 the Hone residence was converted for use by the American Hotel.

considered one of the finest in the city (Fig. 5). Fourteen years later he realized $60,000 from the sale of that house, which, like its neighbors, was to be converted for use by the American Hotel. Entries in Hone's diary describe the accelerated building activity of the 1830s, triggered in part by the disastrous fire of 1835. "The pulling down of houses and stores in the lower parts [of the city] is awful," he noted in 1839: "Brickbats, rafters, and slates are showering down in every direction. There is no safety of the sidewalks, and the head must be saved at the expense of dirtying the boots. The spirit of pulling down and building up is abroad. The whole of New York is rebuilt about once in ten years." Ten years later the situation had not altered. One observer warned strangers that in New York "nothing is fixed, nothing is permanently settled—all is moving and removing, organizing and disorganizing, building up and tearing down; the ever active spirit of change seems to pervade all bodies, all things and all places in this mighty metropolis." [8]

⁸ Cooper, *New York, Being an Introduction to an Unpublished Manuscript, by the Author, Entitled the Towns of*

While this continual destruction was very much a function of business expansion, some observers felt that changes in fashion played a major role and that merchant princes were forever "pulling down and building up, with the same matter of fact indifference, as though changing a spring garment for a summer one." Whether motivated by fashion or economics, new construction was regarded as a harbinger of progress. "Architecture must, of necessity, progress among a people so migratory as ours," wrote one critic who detailed the sequence of events.

*Manhattan* (New York: William Farquhar Payson, 1930), pp. 11–16; Dickens quoted in Logan, "That Was New York," p. 86; Allan Nevins, ed., *The Diary of Philip Hone 1828–1851* (2d ed.; New York: Dodd, Mead & Co., 1936), pp. 201–2; John C. Myers, *Sketches on a Tour through the Northern and Eastern States, the Canadas and Nova Scotia* (Harrisonburg, Va., 1849), pp. 50–51, quoted in Kramer, "Contemporary Descriptions," p. 267. For discussions of the Hone house, see Berry B. Tracy, "For 'One of the Most Genteel Residences in the City,'" *Bulletin of the Metropolitan Museum of Art* 25, no. 8 (Apr. 1967): 283–91; Robert Nikirk, "Philip Hone, Auctioneer-Gentleman-Mayor," *Auction* 2, no. 4 (Dec. 1968): 7–9.

Ever changing, ever new. The business locations of our cities are ever encroaching on the sphere of fashionable residents; and the latter, affrighted, are ever on the alert to fly to regions more remote from the vulgar ways of trade. Thus it is with what, a short time ago, were palatial residences. The leaders of society have gone thence forever; the merchant and the boarding-house keeper have taken their places; and the architectural front, that once was the pride of its owner, would now be looked at by him with a feeling of surprise, to think how mean his ideas once were.[9]

Our observer overlooked one important group of people—the urban poor. The population of New York was swollen by an influx of immigrants who occupied the abandoned older portions of the city or who built makeshift shanties on barren lots or in the still-undeveloped northern regions of the island. New York only gradually filled out the grid that had been laid out in 1811, and the settlement pattern fanned out slowly into Brooklyn, Williamsburg, Westchester, Staten Island, and New Jersey. The poor often lived cheek by jowl with the wealthy, and the coexistence of squalor and elegance seemed somehow inevitable in a city in which there were "obvious and actual extremes of fortune, character, violence, philanthropy, indifference and zeal, taste and vulgarity, isolation and gregariousness, business and pleasure, vice and piety." The poor became a polarizing element, and the contrasts of poverty and wealth not only filled the popular literature but also provided a cause for increasing humanitarian enterprise (Fig. 6). In addition to campaigns for land-use reform and pleas for the creation of useful urban amenities, a peculiar stimulus to the expression of private wealth emerged: it was increasingly stated that if the rich applied their fortunes to noble, edifying, or tasteful enterprises, the poor would learn by example.[10]

Objections have been made, on moral and economical grounds, to the display of wealth and splendor in archi-

Fig. 6. Housing for the rich and housing for the poor in New York City. From Matthew Hale Smith, *Sunshine and Shadow in New York* (Hartford, Conn.: J. B. Burr & Co., 1868), frontispiece.

[9] "Progress of Architecture in the United States," *Sloan's Architectural Review and American Builders' Journal* 1, no. 4 (Oct. 1868): 283–84.

[10] John W. Francis, *Old New York; or, Reminiscences of the Past Sixty Years* (1865; reprint ed., New York: Benjamin Blom, 1971), p. 378. For artistic responses to emerging social institutions, see Neil Harris, *The Artist in American Society, The Formative Years 1790–1860* (New York: George Braziller, 1966); Albert Fine, "The American City: The Ideal and the Real," in *The Rise of an American Architecture,* ed. Edgar Kaufmann, Jr. (New York: Praeger Publishers and The Metropolitan Museum of Art, 1970), pp. 51–59.

tectural decorations, but, we cannot think with justice. We regard it as the mere natural and normal expression of progress, the counterpart of that formerly exhibited in the great commercial republics of Italy and Holland. Luxury is a vice, only when it is extravagance in an individual: the private vices of ostentation and extravagance become public benefits to trade and industry. The due scale of expense for every grade of society can never be fixed by lawgiver or moralist. The sumptuous environments of the richest merchant are by use and familiarity no greater luxuries to him, than more homely comforts are to the mechanic; and in a country, where all are striving to get rich, it may seem hypocrisy and envy, to cavil at the use and display of riches. But, viewed in a public light, every external indication of prosperity tends to add attractions to a city, and to promote its increase and influence in more important objects.

The preceding passage, which appeared in 1854 in *Putnam's Monthly*, is quoted at length as it effectively summarizes the charges and responses that circulated around the millionaire's mansion for the following fifty years. It provides a significant a priori justification of opulent display as a social and economic expedient. The *Putnam's* critic went so far as to suggest that "we ought to feel grateful to these men who are willing to lavish their wealth in the erection of costly houses which so beautify our streets and thoroughfares, and render a walk through our avenues as agreeable as a visit to a gallery of art." Nor could he find such extravagance objectionable regardless of the motivation. "We will not quarrel with those who contribute in any manner to the public welfare, even though in doing so they have no higher object than self-glorification." Thus, at a time when slums were expanding and more and more people were living in tenements and boardinghouses, the construction of palatial private homes found a curious rationale.[11]

Much of the housing demand was met by uncommissioned, speculatively built structures, and even the wealthy were often content to live in these or in moderately pretentious houses. Such conditions prompted one English critic to quip, "I should say that America must be the paradise of builders and the purgatory of architectural connoisseurs." The number of new buildings erected annually in New York more than doubled between 1834 and 1849. In addition to the expanding population, an increasing number of visitors thronged the streets and avenues, attracted to the shops and the showplaces and that captivating commingling of the two, New York's Crystal Palace (1853–58). New hotels and houses, shops, and places of amusement were required to meet the demands of this increased traffic, and the prevailing architectural style of these buildings was the Italianate, derived largely from English design sources.[12]

The Italianate style was an amalgam of formal and pictorial aspects of seemingly opposed styles: the medieval and the classical. Its appeal lay in its adaptability and in the impression of modernity that it created. In suburban and rural areas, Italian villas, with their picturesque broken outlines and asymmetrical massings, were considered harmonious with the landscape, unlike the more severely classical structures, which appeared aloof and uninflected. In the city, Italianate-style buildings were regularized in their distribution of details and made symmetrical in their arrangement of features, but were elaborated through ornamental detailing. The Italianate, based as it was on domestic architectural models rather than on temple forms, permitted increases in size required by the extended facades of urban buildings without introducing incongruities in style. Its desirable qualities were effectively summarized by one contemporary: "The Italo-Roman, or Palladian style, should be confined to blocks of houses, stores, or other continuous ranges of tenements, in the streets of cities, where numerous stories, windows, and doors are indispensable, as they can be made to harmonize with that manner of construction, and the exuberance of ornaments which it admits, will have a pleasing effect, by varying the monotonous extent of prolonged surfaces." The quest for greater picturesqueness and more expressive forms had a profound effect on both rural and urban construction. "The deserted followers of Greek and Roman forms," wrote one retrospective observer in 1868, "seeing shrewdly where their interest pointed, laid Benjamin and Lafever on the topmost shelf, and owned that there was something in the new 'fangle.' The progress of architectural taste became now very decided and professional architects were sought after and well patronized. The stimulus had its due effect, and the appearance of our buildings, private and public, was decidedly improved."[13]

New York in the 1850s was a city of stores, hotels, and private houses cast in a Mediterranean mold. The elaboration of moldings at windows, doors, and cornices combined with rusticated basements

---

[11] Clarence Cook, "New York Daguerreotyped—Private Residences," *Putnam's Monthly* 3, no. 15 (Mar. 1854): 241, 247. For an interesting discussion of some reasons for and reactions to the great private house in the later nineteenth century, see Edward C. Kirkland, *Dream and Thought in the Business Community* (Ithaca: Cornell University Press, 1956), chap. 2, "The Big House."

[12] Anonymous critic quoted in Louisa Caroline Tuthill, *History of Architecture from the Earliest Times; Its Present Condition in Europe and the United States . . .* (Philadelphia: Lindsay & Blakiston, 1848), p. 301.

[13] "The Arts," *American Quarterly Register and Magazine* 1 (May 1848): 188; "Progress of Architecture," p. 278. For a contemporary defense of the Italianate, see E. M. Field, *City Architecture: or Designs for Dwelling Houses, Stores, Hotels, etc.* (New York: G. P. Putnam & Co., 1853). See also Charles Lockwood, "The Italianate Dwelling House in New York City," *Journal of the Society of Architectural Historians* 31, no. 2 (May 1972): 145–51.

FIG. 7. Looking south on Fifth Avenue, New York City, 1859. Detail from "New York City and Environs from the Spire at Dr. Spring's New Brick Church," *Harper's Weekly* 3, no. 121 (Apr. 23, 1859): 264–65.

and quoins to give streets greater variety and modulation. The increased scale of houses, the larger windows, and the deeper punctures of openings into newly popular brownstone fronts provided textural qualities that the less effusive Greek-style houses had lacked. New York perfected its own version of the Italianate house with the appropriation of the high Dutch stoop, and this urban vernacular, with some ornamental variations, continued in currency throughout much of the century. But even these elaborations, when repeated over continuous ranges of buildings and duplicated through long vistas of regimented avenues and side streets, came to be regarded as a plague of brownstone nonentities lacking true monumentality and distinction (Fig. 7). "What artist, coming here from a distance in search of architectural knowledge," inquired a critic for the *New York World,*

needs any sketch-book in traversing our much vaunted Fifth avenue, from end to end? When he has seen one house he has seen them all. The same everlasting high stoops and gloomy brown-stone fronts; the same number of holes punched in precisely the same place, and only ringing the changes upon square, circular or segmental heads; the same huge cornices bristling with overpowering consoles and projections, and often looking, in their cumbersome and exaggerated proportions, like whole regiments of petrified buffaloes leaping headlong from the roof. To be sure, what we lack in invention we can cover over by "ornamentation." [14]

In contrast to this diatribe, individual mansions were frequently described in the most extravagant terms. One of the last great houses in the Greek revival style was constructed in 1845–47 for John

[14] "Architecture in New York," *New York World*, quoted in John W. Kennion, *The Architect's and Builder's Guide* (New York: Fitzpatrick & Hunter, 1868), p. 33.

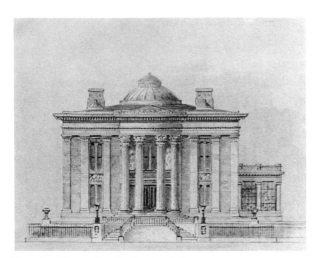

FIG. 8. Alexander J. Davis, John Cox Stevens house. New York City, 1845–47. Watercolor. (Avery Architectural Library, Columbia University.)

Cox Stevens at College Place, after designs by Alexander J. Davis (Fig. 8). This Corinthian villa prompted the diarist Philip Hone to extol, "The palais Bourbon in Paris, Buckingham Palace in London and the Sans Souci at Berlin are little grander than this residence of a simple citizen of our Republican city." Stevens's elegant marble mansion stood for only seven years before the tide of commerce obliterated it. Most of the new generation of mansions appearing at the same time were in the Italianate manner or in its variant, the Anglo-Italian palazzo style. Thus, as Stevens's house was being built, one of the earliest freestanding Italianate houses was also under construction, a mansion for Col. Herman Thorne at 22 West Sixteenth Street (Fig. 9). It is not surprising to find the Thorne house described in terms of its contribution to the metropolitan scene. Calling the house "unquestionably the finest private dwelling in the country," the *United States Magazine, and Democratic Review* noted, "It has an air of unostentatious magnificence that no town house in the Union can pretend to," and concluded, "We are glad for the sake of our domestic architecture, that Col. Thorn[e] has erected a noble mansion in 16th street. It will help vastly to ornament our growing town and to improve the taste and knowledge of our builders." Other fine residences were constructed in the Anglo-Italian palazzo style along Fifth Avenue, now a fashionable rival to increasingly commercial Broadway. Contemporary descriptions of the new elegances of the city abounded

in newspapers and journals, and a spirit of pride in achievement and optimism for the future prompted ambitious claims. A more balanced appraisal was offered by the *Putnam's* critic who wrote:

It is in the private mansions which are built, ornamented and furnished to conform to the tastes, the incomes, and the exigencies of their occupants and not in the public edifices that we must look for the true development of the national taste. . . . Hence we have an immense number of very fine houses which, in the aggregate, form streets of greater beauty than any city of the old world can boast of, but no single building to be compared with the splendid triumphs of architecture which constitute the glory and attraction of Paris.[15]

Paris increasingly became the urban model for New York, and although the full Second Empire

[15] Nevins, *Diary of Philip Hone*, pp. 868–69; "Our New Houses," *United States Magazine, and Democratic Review* 21, no. 113 (Nov. 1847): 393–94, quoted in Lockwood, "Italianate Dwelling House," p. 148; Cook, "New York Daguerreotyped—Private Residences," p. 233. For fuller discussion of New York houses in this period, see Charles Lockwood, *Bricks and Brownstone, The New York Row House, 1783–1929, An Architectural and Social History* (New York: McGraw-Hill Book Co., 1972); Ellen Kramer, "The Domestic Architecture of Detlef Liefau, A Conservative Victorian" (Ph.D. diss., New York University, 1957).

FIG. 9. Col. Herman Thorne house, New York City, 1846–48. From Clarence Cook, "New York Daguerreotyped—Private Residences," *Putnam's Monthly* 3, no. 15 (Mar. 1854): 242.

FIG. 10. Looking north on Fifth Avenue from Thirty-third Street, New York City, ca. 1877. *Foreground,* John Jacob Astor II house, 1859; *middle,* William B. Astor II house, 1856; *far right,* A. T. Stewart house, 1864–69. (New-York Historical Society.)

style did not arrive until the 1860s, occasional couplings of Italianate and French elements took place. Usually these consisted of grafting a French mansard roof onto an essentially Italianate building. With the appearance of the mansard, the principal components of New York's urban architectural style were established for the next quarter century.

Two houses built for members of the Astor family during the 1850s provide interesting examples of the grand but restrained manner still favored by the wealthy. While William B. Astor, the son of John Jacob Astor and the city's wealthiest citizen, lived in a house on Astor Place given to him by his father, his sons sought ampler accommodations on Fifth Avenue. John Jacob II and William B. II constructed houses considerably further north, on the west side of the avenue, at the corners of Thirty-third and Thirty-fourth streets, respectively (Fig. 10). William B.'s house was built in 1856. A simple brownstone-front mansion four stories high above a basement, its architectural embellishments were limited to modest projections above the windows. Triangular pediments crowning the first-

floor windows and door provided the only elaboration. The slightly later (1859) house that John Jacob built on the northwest corner of Thirty-third street was more elegant. A three-story structure of Philadelphia pressed brick with brownstone trim, it was topped by a mansard roof. The house was set back from the street and raised on a basement punctured by round arched windows. The facades were articulated by belt courses, with the course separating the first and second floors elaborated to form a balustrade in front of the second-floor windows. These windows were capped by triangular pediments. A pavilion arrangement indicated by a slight forward break in the facade and a quoinlike framing to the pavilion bays created a vertical emphasis. While the flat roof of the William B. Astor house placed great emphasis on the facade, the mansard of John Jacob's house suggested a volumetric quality, further enforced by the rounding off of corners, a treatment recalled in notches at the corners of the roof. Both buildings displayed blank walls on the inner sides, which faced each other across a garden. Irregularly introduced windows provided the only relief, although these walls were

FIG. 11. Samuel P. Townsend house, New York City, 1854–55. Photograph, ca. 1860. (Museum of the City of New York.)

clearly visible from the street. It is possible that this was done with the expectation that another house would occupy the intervening space, but the houses remained isolated until they were demolished.

Across Thirty-fourth Street the Astors had a parvenu neighbor, the sarsaparilla patentee, Dr. Samuel P. Townsend, who had erected a blocky brownstone pile in 1854–55 (Fig. 11). The Townsend house, less sophisticated in detail than the Astor houses, featured enormous windows capped by flattened pediments, awkwardly introduced balconies, and overhanging bay windows. With a thin projecting flat roof and multifaceted monitor, the house had a quixotic air that was well described by one commentator: "It was the largest in New York, built of brown stone, as gorgeous and inconvenient as an Eastern pagoda. It cost fabulous sums. It was large enough for a hotel, and showy enough for a prince. It was burnished with gold and silver, and elaborately ornamented with costly paintings. It was the nine days' wonder in the city, and men and women crowded to see it at twenty-five cents a head. The owner failed, and the house passed out of his hands. It became a school, with no success." Within ten years of its construction, the Townsend house was demolished to make way for a grander, showier, and infinitely more expensive palace. George Templeton Strong, diarist and member of the Knickerbocker establishment, noted the transi-

tion in a diary entry of March 21, 1864. "Up Fifth Avenue this morning to call on Mrs. William Astor, and get instructions for drawing her will. . . . The great, hideous one-hundred-thousand-dollar Townsend-Sarsaparilla-Springler house on the other side of Thirty-fourth street has just been bought by A. T. Stewart, who has razed it to the ground and tells William Astor he is going to lay out one million on a new white marble palazzo. I suppose it will be just ten times as ugly and barbaric as its predecessor, if that be conceivable." [16]

Strong's foreboding about Stewart's impending enterprise was probably rooted in his antagonism toward pretension. Yet up to this time Stewart had lived in DePau Row, a modest terrace block on Bleecker Street (Fig. 12). Stewart had, however, built prominently and had already made an enormous impact on the face of the city. The marble palace, now nearly twenty years old, was built in the emerging Italianate vocabulary. His use of rich materials, daring structure, and good design had made the architecture of the building synonymous with his business. The cast iron retail store, built in 1859 further up the island on Broadway just below fashionable Grace Church, eventually occupied the entire block from Ninth to Tenth streets and from Broadway to Fourth Avenue (Fig. 13). When completed, it was the largest cast iron structure in New York and one of the most prominent buildings in this novel material.[17]

The designer of Stewart's cast iron store was John Kellum (1809–76), who authored all subsequent Stewart projects in the New York area, including the mansion, a hotel for workingwomen, and the original domestic and public buildings of Garden City, a planned community Stewart began developing on Hempstead Plains, a large tract of land in Long Island that he acquired in 1869. Kellum was, in fact, a native of the town of Hemp-

[16] Smith, *Sunshine and Shadow*, p. 61; Allan Nevins and Milton Halsey Thomas, eds., *The Diary of George Templeton Strong, 1835–1875*, 4 vols. (New York: Macmillan Co., 1952), 3:416.

[17] Cook, "New York Daguerreotyped," *Putnam's Monthly* 1, no. 4 (Apr. 1853): 358, 362; Resseguie, "Marble Palace," pp. 154–62; Kramer, "Contemporary Descriptions," pp. 270–71; Carl Condit, *American Building Art, The Nineteenth Century* (New York: Oxford University Press, 1960), pp. 41–42; Alan Burnham, "Last Look at a Structural Landmark; John Wanamaker's Store," *Architectural Record* 120, no. 3 (Sept. 1956): 273–79; *A History of Real Estate Building and Architecture in New York City during the Last Quarter of a Century* (New York: Record & Guide, 1898), pp. 459, 461–62.

FIG. 12. DePau Row, south side of Bleecker Street, from Thompson to Sullivan streets, New York City, 1896. (New-York Historical Society.)

FIG. 13. John Kellum, A. T. Stewart's cast iron store. New York City, 1859. From *Architectural Record* 1, no. 2 (Oct.–Dec. 1891): 246, showing cast iron store when it occupied the block from Broadway to Fourth Avenue, and from Ninth to Tenth streets.

stead and had been a carpenter's apprentice there before moving to Brooklyn, where he first appeared in directories as a carpenter in 1842–43. By 1850 he had formed an architectural partnership with Gamaliel King. In 1855 they opened a branch office in New York and at about the same time designed a store for Wilson G. Hunt, which, according to Richard Lathers, brought Kellum to Stewart's attention. It may well have been Stewart's commission that precipitated the dissolution of the King and Kellum partnership and the formation, probably in 1858, of the firm of Kellum and Son.[18]

In Stewart's cast iron store, Kellum adapted to the exigencies of construction techniques and responded with the loosely Renaissance style much in favor for cast iron buildings. Deeply set arched openings framed by engaged columns provided an articulated bay unit that could with grace and logic be extended through the long facades and many stories of the building. Although intended from the beginning to occupy the entire block, the store was built in stages. The precision of continuous arcading lost its architectural logic in the overall ensemble because the bay units, grouped between rusticated piers, were of unequal lengths. Kellum's main failure was in the organization and articulation of the larger shapes and units and the subordination of detail areas.

The cast iron store, like its downtown cousin (now devoted to Stewart's wholesale operation), was a major addition to the New York scene when it opened for business in 1862. Like the marble palace, it was organized around a central well, and the scale of the enterprise was readily apparent. The exterior was painted white in emulation of masonry, a common practice at the time. Stewart is said to have "likened his iron front to puffs of

white clouds, arch upon arch, rising 85 feet above the sidewalk," but others were more critical. One commentator found the brilliant white facade offensive and "without any relief whatever except the ghastly one of the blue cotton window shades." He remarked that "on a fine day, when these shades are all pulled down, the architect of this building, if, indeed, it ever had one, must be congratulated upon having manufactured the most purely ugly and conspicuously offensive structure in New York City if not in the whole continent." But by the time these comments were made, Kellum was at work on the costliest and most luxurious mansion New York had ever seen.[19]

A. T. Stewart's marble mansion (1864–69) was calculated to impress, and the builder's choice of a site directly adjacent to the Astor houses on Fifth Avenue was certainly deliberate (Fig. 14). The purchase of the Townsend property gave Stewart an unusually large lot. While the average New York City lot was 25 feet wide and 100 feet deep, Stewart's property extended 150 feet on Thirty-fourth Street and 111.9 feet on Fifth Avenue. The merchant "found it brownstone and left it marble." Although he first intended to remodel the Townsend house, and went as far as gutting the structure, by the time of George Templeton Strong's visit to the Astors he had entirely demolished the old fabric and had begun construction of a mansarded palazzo. In addition to the size of the lot and the scale of the house, the style served to distinguish the mansion from its predecessors. Although the French style had arrived in the United States shortly after the first Italianate buildings were constructed, it had not won immediate acceptance. The "mansard mania" that dominated the late 1860s and 1870s was just beginning when Stewart's house was under construction. Whether Stewart or Kellum was responsible for the choice of the Second Empire (or modern French) style is difficult to determine. Outside of his projects for Stewart, Kellum was known primarily for public buildings, many in cast iron, for which custom had already declared the Renaissance style the most suitable. An 1850s advertisement for the firm of King and Kellum included an illustration of a typical car-

[18] *Appleton's Cyclopedia of American Biography* (1888), s.v. "Kellum, John"; "Two Popular Architects," A. J. Bloor to *American Architect and Building News* 5, no. 166 (Mar. 1879): 71; "Correspondence," *American Architect and Buildings news* 1, no. [26] (June 24, 1876): 206–7; Alvan Sanborn, ed., *Reminiscences of Richard Lathers, Sixty Years of a Busy Life in South Carolina, Massachusetts and New York* ([New York]: Grafton Press, 1907), p. 42. Resseguie, "Decline and Fall," p. 261, n. 28, points out that Lathers was an old man indeed when the *Reminiscences* were written and that his information about Stewart is not always to be trusted. For Garden City, see Mildred H. Smith, *History of Garden City* (Manhasset, N.Y.: Channel Press, 1963); Roger A. Wines, "A. T. Stewart and Garden City," *Nassau County Historical Journal* 19, no. 1 (Winter 1958): 1–15.

[19] "Men Who Have Assisted in the Development of Architectural Resources—No. 1. John B. Cornell," *Architectural Record* 1, no. 2 (Oct.–Dec. 1891): 245; "Our New York Letter," *American Builder and Journal of Art* 1, no. [4] (Feb. 1869): 49.

FIG. 14. John Kellum, A. T. Stewart house. New York City, 1864–69. Photograph, ca. 1885. (Museum of the City of New York.)

penter Gothic cottage, but this was probably no more than a standard typographical cut and not necessarily indicative of the firm's product. It does, however, suggest that they wished to be known as domestic architects. An anecdote recounted by Richard Lathers implies that at least in scale and pretension the specifications were Stewart's. According to Lathers, when the architect first showed his designs to Stewart, the merchant demanded, "Is that the best you can do for a fine mansion?" The architect replied, "Oh, not at all. I could design a palace if you were willing to build it." Stewart had twice built in an emerging style and, whether by chance or intention, was associated with architectural innovation.[20]

[20] Sanborn, *Reminiscences of Lathers,* p. 43. See *Atlas of the City of New York Borough of Manhattan,* vol. 2, *14th Street to 59th Street from Actual Surveys and Official Plans*

by *George W. and Walter S. Bromley* (New York: G. W. Bromley & Co., 1899), pl. 21, block 836, lots 33–39; *New York Times,* Apr. 11, 12, 1876; Conveyance Liber 869, Dec. 6, 1862, p. 96. At a sheriff's sale on October 17, 1862, Stewart bought the Townsend property, a lot 98′8″ on Fifth Avenue by 150′ on Thirty-fourth Street, for $73,500 Conveyance Liber 875, June 10, 1863, p. 681. On May 30, 1863, Stewart bought an additional parcel adjoining his northern boundary 13′ by 150′ from John and Mary H. Caswell for $13,500. At the time of Stewart's death the marble mansion was appraised for taxes at $500,000, and the older house at 355 Fifth Avenue, into which Stewart had moved during the construction of the mansion, at $70,000. It was noted that the appraisals were at 60 percent of value. See Montgomery Schuyler, "A Modern Classic," *Architectural Record* 15, no. 5 (May 1904): 431–44, reprinted in Schuyler, *American Architecture and Other Writings,* ed. William H. Jordy and Ralph Coe, 2 vols. (Cambridge: Harvard University Press, Belknap Press, 1961), 2:588, which discusses McKim, Mead, and White's Knickerbocker Trust Company, 1902–4, the structure that replaced the Stewart mansion. For a list of Kellum's cast iron projects, see

Fig. 15. John Kellum, A. T. Stewart house. New York City, 1864–69. Photograph, ca. 1885, showing the awkward corner arrangement of pilasters on the western side of the Thirty-fourth Street facade. (Picture Collection, New York Public Library.)

The Stewart house was more monumental and more picturesque than its predecessors. While respecting the grid plan of the city, it was set off from the street and was meant to be seen as a three-dimensional entity. Unlike the neighboring Astor houses, it was fully detailed on all sides. Formally, the house derived its style from masonry construction (Fig. 15). Rising three stories above a basement, it was crowned by a mansard roof with projecting dormer windows. Although the house was

six bays wide fronting on Thirty-fourth Street, the facade was broken into a pavilion arrangement encompassing all but the bay closest to Fifth Avenue. This clustering of the five western units allowed a symmetrical arrangement of bays on either side of the entry. A further slight projection from the line of the facade marked the central bay, which was reached by a broad flight of steps reported to have been thirty feet wide with each stair hewn from a single slab of marble. The entry featured a peculiar porch arrangement of paired columns bearing a full entablature at the first story. This, in turn, was surmounted by a second pair of columns capped by entablature blocks supporting scrolled volutes that abutted the walls on either side of the third-story window. While the entablature of the first level of the porch was continuous with an entablature

entries for "King and Kellum" and "Kellum and Son" in Daniel D. Badger, *Architectural Iron Works of New York— Illustrations of Iron Architecture* (New York: Daniel D. Badger, 1865). Kennion, *Architect's and Builder's Guide,* pp. 38–40, provides the best description of construction details of the mansion.

Fig. 16. Hall of the A. T. Stewart house, New York City, ca. 1880s. From *Artistic Houses,* 2 vols. (New York: D. Appleton & Co., 1883–84), 1, pt. 1: opp. 8.

course encircling the building, the second-floor entablature blocks terminated at the wall. A narrow course provided demarcation between the second and third floors. This eccentric columnar arrangement was repeated on the central of the three Fifth Avenue bays. The western face of the building, opposite the Fifth Avenue front, had four bays, thus defeating attempts at providing a central motif. Here the columnar porch was attached to the bay closest to Thirty-fourth Street, complicating the detailing at the corner.

The first-floor bay units were indicated by paired pilasters, while single rusticated pilasters served the same function on the second and third stories. These pilasters were doubled on the third floor above the entry and in the center of the Fifth Avenue facade. Only the plain pilasters of the first story had ornamental capitals, in the Corinthian order. The Fifth Avenue corner at the first story was marked by a cumbersome arrangement of paired pilasters facing the street and a third turning the corner. Between the pilaster groupings were large rectangular windows surrounded by modest moldings. The second-story windows on either side of the central bay and on the Fifth Avenue side were topped by entablatures recalling the entablature blocks of the columns at that level. Balustraded balconies at the first-floor windows provided a visual counterpoint to the balustrade surrounding the base of the roof.

A simplified entablature with molded architrave, plain frieze, and modillioned cornice encircled the building and was punctuated by large single and double brackets above the wall pilasters. The brackets provided visual support for the cast iron balustrade, which was painted to harmonize with the marble walls. Ornamental detailing along the ridges of the roof provided a final decorative fillip. The contrast between wall color and roof was noticed by one reporter who commented, "Among the many private dwellings of the first class, which have expensive roofs of this formation [mansard], that of A. T. Stewart, the merchant prince, stands prominently forward; and is, in the pleasing tints of its slating a relief to the eye, from the mass of white marble it surmounts." [21]

Attached to the northern side of the house and extending beyond the western wall was a lower block, nearly two stories high, which contained Stewart's art gallery. This unit was about 80 feet long and 30 feet wide. The main body of the house was approximately 120 feet by 72½ feet. The simple interior plan of the main block repeated the five-and-one division of the Thirty-fourth Street facade. The entrance gave onto an impressive hall cut at midpoint by a transverse corridor (Fig. 16),

[21] "Domestic Architecture," *Sloan's Architectural Review and American Builders' Journal* 1, no. [4] (Oct. 1868): 254.

which divided the space of the five-bay unit into four principal rooms: a music room (Fig. 17), a reception room (Fig. 18), a dining room, and a breakfast room, which was somewhat diminished by the adjacent staircase. At the Fifth Avenue end of the transverse corridor, and parallel to the main hall, a large drawing room that corresponded to the single bay on the exterior extended the full width of the house (Fig. 19). The library occupied the same position on the second floor (Fig. 20). In addition to the library, the second floor contained the principal bedrooms (Fig. 21), with an adjacent dressing room, a sitting room, and a billiard room. The second and third floors were each divided into eight rooms, with a single major room occupying the easterly bay. Above the third story was an entresol divided into eight rooms, each 6 feet high. Two additional floors were accommodated under the roof, a fourth floor with rooms 9½ feet high and an attic floor of 6-foot-high rooms. With the rooms on the first two floors well over 18 feet high, the house achieved monumental proportions. Buried within the marble walls was an extensive system of iron supporting elements. The principal metal structural elements were floor beams, lathing, rafters, and a corrugated iron roof, but iron was also employed in auxiliary ways, especially for piping, water storage tanks, and the stairs. Brick arch construction resting on iron beams supported the weight of the marble and tile floors and guaranteed fireproof construction.

The interior was sculptural in detail and commanding in effect. Window and door surrounds were cased with marble framing carved in Italy, and the walls of the entrance hall were paneled with marble. This hall, with its freestanding fluted Corinthian columns at the junction with the cross hall, became a triumphal entryway, quickly establishing the richness of the entire house. Giant mirrors reflected the light given off from enormous gas torchères. Monumental sculptures by the Italians Antonio Tantardini and Adam Scipione Tadolini and the Americans Harriet Hosmer, Randolph Rogers, and Thomas Crawford peopled the hall in mute testimony to the owner's cultural attainments. These figures vied for attention with "an immense French clock, from the factory of Eugène Cornu, Paris, surmounted by a silvered bronze figure, holding in her right hand a swaying pendulum, the whole fourteen feet high, and indicating not only the hour, minute, and second of the day, but also the day of the week, the change of the moon, the record of the barometer and thermometer, and various other matters." [22]

To some contemporaries the house appeared a monument as much to art as to trade. With ready pens they traced exultant tributes to its splendor.

The building, with scarcely an alteration in the arrangement of its rooms, could be transformed into a magnificent art-gallery. It almost astonishes us to hear the architecture speak of this as a reception room, of that as a breakfast room, and of another as the parlor. The beautiful wardrobe and bath-rooms are the only portions of the house which distinctively suggest the idea of a private residence. The vista of rooms is one of the most remarkable features which strike the eye. From one end of the building one looks through to the other. The grand hall leads to the marble staircase—the most beautiful specimen of architecture of that kind in the country. Passing beyond this, we enter the picture-gallery, which is to be adorned with the finest specimens of statuary and painting that *Mr. Stewart* has been able to obtain during a search of several years.[23]

The picture gallery in the lower block extended nearly the full length of the house. This large sky-lit room had a coved ceiling ornamented with trompe l'oeil painting suggesting a running balcony of alternating bowed pedestals and broken scroll volutes (Fig. 22). The painted pedestals were topped with illusionistic sculptured busts; above the broken scrolls, portrait heads were painted in simulated relief and surrounded by trompe l'oeil frames. The portraits were of prominent contemporary American and European artists. Daniel Huntington, Frederick Church, Albert Bierstadt, and Charles Loring Elliot represented American art, while Rosa Bonheur, Hippolyte Paul Delaroche, Thomas Couture, Horace Vernet, Jean Léon Gérôme, and Jean Louis Meissonier were chosen for the French school. Actually Stewart had not patronized American artists liberally, and his collection leaned heavily toward contemporary French academic painters—a hint of the penchant for European artists that would dominate the selection of pictures for galleries of wealthy Americans in succeeding decades. Apparently the pictures had been

[22] *Artistic Houses*, 2 vols. (1883; reprint ed., 2 vols. in 1, New York: Benjamin Blom, 1971), 1, pt. 1: 7–8.

[23] "Mr. Stewart's New Residence," *Harper's Weekly* 13, no. 659 (Aug. 14, 1869): 526. See also "The Stewart Art Gallery," *Harper's Weekly* 23, no. 1166 (May 3, 1879): 348–50; *Catalogue of the A. T. Stewart Collection of Paintings, Sculptures, and Other Objects of Art, To Be Sold By Auction . . . March Twenty-Eighth and Following Days . . .* (New York: American Art Assn., 1887).

FIG. 17. Music room of the A. T. Stewart house, New York City, ca. 1880s. From *Artistic Houses,* 2 vols. (New York: D. Appleton & Co., 1883–84), 1, pt. 1: opp. 11.

FIG. 18. Reception room of the A. T. Stewart house, New York City, ca. 1880s. From *Artistic Houses,* 2 vols. (New York: D. Appleton & Co., 1883–84), 1, pt. 1: opp. 9. (Photo, the author.)

FIG. 19. Drawing room of the A. T. Stewart house, New York City, ca. 1880s. From *Artistic Houses,* 2 vols. (New York: D. Appleton & Co., 1883–84), 1, pt. 1: opp. 14.

FIG. 20. Library of the A. T. Stewart house, New York City, ca. 1880s. From *Artistic Houses,* 2 vols. (New York: D. Appleton & Co., 1883–84), 1, pt. 1: opp. 10.

FIG. 21. Bedroom of the A. T. Stewart house, New York City, ca. 1880s. From *Artistic Houses,* 2 vols. (New York: D. Appleton & Co., 1883–84), 1, pt. 1: opp. 17.

FIG. 22. Picture gallery of the A. T. Stewart house, New York City, ca. 1880s. From *Artistic Houses,* 2 vols. (New York: D. Appleton & Co., 1883–84), 1, pt. 1: opp. 15.

acquired in a search of several years, as the bulk of the dated paintings listed in the Stewart sale catalog (1886) fall into the decade 1864–74.

On occasion Stewart commissioned a picture, and the sale catalog entries give an idea of the merchant's aesthetic criteria. Concerning the painting *Return from the Harvest* (1878), by Adolphe William Bouguereau, it noted: "The commission for the above picture was placed with M. Bouguereau by the late Mr. A. T. Stewart in 1874 with the understanding that the painting was to be the artist's greatest work, and not a nude subject; the picture was not finished until after the death of Mr. Stewart. When finished in 1878 M. Bouguereau stated that he considered the work his masterpiece." Other entries in the catalog attest to the cordial relationships Stewart had with artists. Whether this was evidence of clever salesmanship on the part of the artists or a genuine expression of affection and regard, it does suggest that Stewart was personally familiar with European artistic circles. A particularly interesting exchange of letters concerns Meissonier's masterpiece, *Friedland, 1807* (1868), one of the gems of the Stewart collection acquired for a reported price of $60,000. Apparently responding to a request from Stewart, the artist included a self-portrait when he shipped the *Friedland*. Stewart replied, "It gives me pleasure to acknowledge the safe arrival of your superb picture, 'Friedland, 1807,' also your valued letter, with the portrait, as a souvenir of regard; which I shall always cherish and prize as coming '*from my friend Meissonier.*'" Another gem of the collection, the outsized *Horse Fair* (1853), by Rosa Bonheur, was an older picture when Stewart acquired it from William P. Wright of Weehawken, New Jersey. Both pictures were considered major acquisitions when they were given to the Metropolitan Museum of Art, after Mrs. Stewart's death. Although Stewart was among the founding trustees of the Metropolitan Museum, his art collecting seems to have been more calculated than compulsive. The results were summarized by a contemporary who wrote: "As a buyer, Mr. Stewart, who did not profess to be profoundly learned in art, wisely sought the advice and opinion of those who knew something of the merits of the various modern artists and the value of their productions, and thus, except in one or two instances, he got a fair equivalent for his money." [24]

Stewart probably got more than he bargained for with the gigantic canvas *Genius of America* (ca. 1867), by Adolphe Yvon. This 35-by-22-foot celebration of America's bounty was too large even for the marble mansion and had to be exhibited in the ballroom of the Grand Union Hotel in Saratoga, which Stewart had acquired and enlarged in 1872. An appropriate backdrop for the affluent progeny of America's genius, the painting, commissioned by Stewart, was a gigantic allegory of American enterprise. The message was quite clear—America, the land of opportunity, was where Stewart had had such prodigal success. The sale catalog dutifully recorded the sermon: "While the sun is rising on the consummation of the great Centennial, vessels of many nations are bearing hosts of emigrants to the land of freedom, where Industry, Enterprise, Education, with equal laws and full religious toleration, unite to present, as in a visible vista, equality, wealth, and assured social position." [25]

Stewart's house and all its contents told a similar moral tale. It was "an eternal monument which should also be fitted for a temporal habitation." The scale of the furniture and decorations complied with the grandeur of the architectural fabric. In addition to the statuary placed in the picture gallery and the rest of the house, and paintings hung in the principal reception rooms, decorative painting ornamented the walls and ceilings. These encaustic decorations by the Italian Brigaldi, executed "at a cost exceeding 15,000 dollars," had, at the time of one reporter's visit in the early 1880s, just been cleaned with "common soap and water, and rubbed with pumice-stone, in order to remove the blackening caused by the heat of the furnace." This refurbishing was required after less than a decade. The painted decorations varied according to the character of the room, but they consisted largely of garlanded flowers interspersed with putti and classical figures. In the reception and drawing rooms, the carpet and ceiling patterns mirrored each other. The drawing room walls and ceiling were painted with floral festoons that played lacy patterns against the broad architectural handling of windows and doors. Gilt furniture was covered

[24] *Catalogue of the Stewart Collection*, pp. 102, 104; *New York Times*, Apr. 11, 12, 1876; Jerry Patterson, "The A. T.

Stewart Sale of 1887," *Auction* 4, no. 2 (Oct. 1970): 45–49, a flawed account based on a contemporary newspaper description of the sale.

[25] *Catalogue of the Stewart Collection*, p. 105. The painting now belongs to the New York State Education Department, State Education Building, Albany.

with old gold and red satin, and the room was described as having a golden tonality.

The interiors illustrated in *Artistic Houses* (1883–84) show the house some years after Stewart's death, but while it was still occupied by his widow. Although there is evidence that Mrs. Stewart added to the furnishings, the house probably looked much the same during Stewart's lifetime. The illustrations suggest that the furniture was of the best quality and that much of it was in the Renaissance revival style, which was just gaining popularity when the house was completed. It was the richness of materials and expense of the furnishings that most captivated the reporter in *Artistic Houses,* and he singled out such features as two $10,000 Sevres vases in the drawing room. The extensive use of gilt and onyx in furniture, chandeliers, and torchères in this room and elsewhere throughout the house was one of the few details in an otherwise sketchy description. The tone of the passage was set by the introductory remarks: "Furniture and hangings seem to have been obtained without care for cost; or, rather, only the most costly seem to have been selected. Money flowed abundantly during the seven years when this white-marble palace was building for a merchant-prince, and in exchange for it came magnificence, splendor, luxury." [26]

By the time of the publication of *Artistic Houses,* the style of Stewart's furniture and decorations was passing from favor, and the inclusion of the house was a testimony to its significance in spite of its somewhat dated furnishings. Most of the text was devoted to describing Stewart's picture gallery, still considered among the finest in the country. Missing from the description was any reference to the furniture makers or designers who provided Stewart with the furnishings. As it was customary for the author to identify decorators and designers involved in the houses he published, the absence of such information in connection with the Stewart house is mysterious. It could well be that Stewart himself was responsible for major decorating decisions, a reasonable suggestion given the fact that Stewart was a merchant and importer dealing in luxury goods. The same European purchasing agents who bought so well for the shop could have

bought equally well for the house. Stewart himself was the honorary head of the American commission to the International Exposition in Paris in 1867, at the very time the house was being erected. His close identification with his business and his total knowledge of its stock suggest he would not have felt out of place making decorating decisions for his house. While the catalog of the Stewart sale lists the New York decorating firm of Pottier and Stymus as makers of two important tables in the house (lots 987 and 992), the papers of the firm give no evidence that they were responsible for anything beyond supplying an occasional piece of furniture. Stewart's role is at least suggested by a statement in *Artistic Houses* that he was responsible for the design of the rosewood furniture in the dining and breakfast rooms. As these rooms are not illustrated, it is impossible to determine their relationship to the rest of the furnishings. [27]

Many New Yorkers who had anxiously awaited the day when the boards and casings of the house's prolonged construction would be removed were not disappointed when the house was finally unveiled. Despite its architectural flaws, the mansion captivated observers and seemed for the moment to fulfill the city's need for great architectural monuments. The *New York Sun* remarked that Stewart had discharged his public responsibility with the construction of this house and chastised "the late Mr. Astor, who never in his life erected a house, improved a lot, or contributed to the architectural history of the city with an eye to the public good." Stewart's house was seen by contemporaries as just such an endowment, and the peculiar notion was voiced that the creation of a private house was a public act of philanthropy. A reporter for *Harper's Weekly,* addressing a national audience, bemoaned the lack of antiquity or elements of picturesque interest in the city but found an exception in this mansion. "There is," he wrote, "one edifice in New York that, if not swallowed up by an earth-quake, will stand as long as the city remains, and will ever be pointed to as a monument of individual enterprise, of far seeing judgement, and of disinterested philanthropy. There is nothing like it in the world, not even among the palaces of the European nobility." *Palatial* was the only word many writers

[26] "Mr. Stewart's New Residence," p. 525; *Artistic Houses* 1, pt. 1: 7. See also "Inside the Stewart Mansion," Miss Grundy's New York letter to the *Boston Courier,* reprinted in *Real Estate Record* 36, no. 924 (Nov. 28, 1885): 1312.

[27] Pottier and Stymus Papers, Museum of the City of New York. An undated note from Stewart requesting "a comfortable surgical chair" is, according to Margaret Stearns, curator of Decorative Arts, the only Stewart reference in the papers.

Fig. 23. Looking south on Fifth Avenue toward the Stewart house at Thirty-fourth Street and, beyond it, the William B. Astor II and John Jacob Astor II houses, New York City, ca. 1883. (New-York Historical Society.)

could summon to describe the effect of the house as it commanded the street and overshadowed the adjacent Astor properties (Fig. 23). The *Harper's* reporter noted: "The entire structure, external and internal, is destitute of showy ornamentation. The style of beauty adopted is very chaste and severe. It is grand without being heavy; it is fine and elaborate without being fanciful. This will appear even more to be the case when the walls are finished and subdued in color, so as to harmonize with the blue-veined marble."[28]

28 *New York Sun* quoted in Logan, "That Was New York," pp. 82–84; "Mr. Stewart's New Residence," pp. 525–26. Even

The Stewart house was completed just as the architectural profession in America was establishing itself on a firm footing with the founding of the first architectural schools. A professional press emerged likewise and established critical standards that attempted to draw lines of distinction between

before the house was occupied, it was described enthusiastically as a feature of the city. See *Appleton's New York Illustrated* (New York: D. Appleton & Co., 1869), pp. 20–34. It was illustrated in *Valentine's Manual of the Corporation of the City of New York* (New York: Joseph Shannon, 1869), opp. p. 200.

quality and ostentation. Thus one reviewer in the *American Builder* issued a stern warning about the possible influence of a house like the Stewart mansion: "A great many people are influenced by the example of so rich a man, and think that he can't be far out in matters of taste when he has shown so much ability in trade." The critic's chief attack was on the architect, who, he felt, was devoid of any originality: "He has looked over a few books of patterns, and decided that, on the whole, the classic styles are about the neatest, and like to be the cheapest in the long run, and he has very prudently determined to stick by them. They require no planning for their interior arrangements, and no decoration for their exteriors, while, at the same time, they are expensive, and by eating up indefinite sums of money, give large profits to the contractor." [29]

Kellum, with meager professional training and a quick eye for profitable situations, was a ready target for professional censure. His career was summarized by a correspondent to the *American Architect and Building News*: "Beyond the Stewart work and this court-house, it is difficult to recall what Kellum ever did do. From a very poor carpenter's foreman, he suddenly blossomed into an 'architect' and rushed on into a goodly fortune if not into much renown. Blackballed by the New York Chapter of the [American Institute of Architects], he failed to comprehend that receiving fees from mechanics as well as clients might prove a proper ground for rejection." According to the *American Builder,* far from ornamenting the city, Kellum was "neck and neck" with the firm of Renwick and Sands "in the race for the prize to be awarded to whomever shall give architecture in New York, the ugliest and most fatal blow." The critic concluded that Kellum's incompetence—coupled with "a millionaire puzzled to spend his income, with no ideas beyond the shop, no notions, no hobbies, and only a confused sense that he, somehow, owes something to the public in the way of external splendor of living, charity and patronage of the fine arts"—had produced what was termed the "packing-box style." The Stewart house was "a parallelogram of stone set on end, and with a profusion of ornament, bad of its kind and ill-chosen, covering it all over in spots." It lacked even the redeeming quality of a spacious parklike setting and was pronounced

"simply, the most ostentatious and pretending and the most ugly house in New York City." [30]

A few weeks after Stewart's death, the architect Peter B. Wight provided an extended notice of the merchant's architectural activities. Remarking again that Stewart had spent upwards of $6 million in conspicuous building, Wight concluded that this investment had largely been a failure and, as a consequence (and because of the scale and prominence of these projects), could only retard the progress of architecture in New York (the *Real Estate Record* went so far as to call Stewart a public enemy). For Wight, the single exception to Stewart's transgressions had been the marble palace store building. He called the mansion, which had been an object of curiosity during the years of building because of the vicissitudes in its construction, the Fonthill Abbey of America and remarked: "It is hard to conceive that any one surrounded by works of art, as was Mr. Stewart, could have had so little understanding of what constituted a work of architecture." Mrs. Richard Morris Hunt, widow of the architect, recalled that Stewart had once referred to the style of the house as Greek, to which Hunt had replied, "Well Mr. Stewart, it may be Greek to you, I assure you it is 'Greek to me' but I don't think it would deceive the smallest little yellow dog that runs down the street." But Wight found the picture not totally black and thought that Stewart's activities might serve a positive purpose.

They may be a warning to the rising generation of architects and great capitalists of the future. Such errors of mis-directed capital are not likely to be repeated; for the community is advancing in taste, discrimination, and common sense, as applied to building improvements. . . . as the general public grows in appreciation, it will insist that capitalists who have it in their power by improving or defacing the public streets of great cities, to advance or retard the progress of intelligence, still be mindful of the onward march of civilization, or else be held up to public criticism.[31]

[29] "Our New York Letter," *American Builder and Journal of Art* 1, no. [5] (Mar., June 1869): 69, 119.

[30] "Correspondence." *American Architect*, pp. 206–7; "Our New York Letter," p. 119. See also the criticism of Kellum in "A. T. Stewart as a Real Estate Operator," *Real Estate Record* 17, no. 420 (Apr. 1, 1876): 237.

[31] Catherine Howland Hunt, comp., Richard Morris Hunt Papers, in the possession of Alan Burnham, New York City Landmarks Preservation Commission, p. 287; Peter B. Wight, "A Millionaire's Architectural Investment," *American Architect and Building News* 1, no. [19] (May 6, 1876): 147–49; "Stewart as a Real Estate Operator," p. 237.

The Stewart mansion had little immediate impact, but it provided a municipal legacy. It was significant, not in terms of style, since it was created just before a panic and depression that forestalled the ultimate development of the millionaire's mansion for a decade, but as a phenomenon. It was the fruition of the impulse to create distinctive private mansions that would enliven the barren vistas and dull conformity of the emergent city. Twenty years after the Stewart house was begun, Mariana G. van Rensselaer appraised its importance in relation to more recent domestic architecture.

The great marble house on the north-west corner of Fifth Avenue and Thirty-fourth street was one of our earliest attempts at novelty, and in ambition it has certainly not since been surpassed. But it was not really a new departure—it was merely an effort to glorify the "vernacular" by increase of size, by isolation, and by change of material. In the last-named respect the effort was commendable. . . . here we have no good proportioning and no skillfull composition either with masses or with features. Beauty has been sought only in the applied columnar decoration, and this is not architecturally valuable because it has been used without moderation, without care for contrast or relief or structural subordination, and without artistic knowledge in design or artistic grace in execution. We can only call it a very showy house, and add that to some eyes it may seem imposing—may seem to deserve the epithet "palatial," which epithet, I imagine, it was the first New York home to suggest to the reportorial pen.

Thus, for the critical contemporary, the Stewart house was a point of departure, a negative reference suggesting the need for greater subtlety and sophistication. At the time van Rensselaer was writing, a new generation of mansions had emerged —the product of trained beaux arts architects, such as Richard Morris Hunt and McKim, Mead, and White. In praising the work of these men, she noted the advances that had been made. The time was finally right for the blossoming of architecture, and she concluded: "It is well that we should see that the richest elaboration need not be ostentatious, much less vulgar; that lavish art may be as refined as modest art; that excess means wrong work." By 1904 the epitaph for the Stewart house had been written. Having served as the Manhattan Club from 1890 to 1899, it was demolished (1902–04) to make way for McKim, Mead, and White's Knickerbocker Trust Company. Montgomery Schuyler, praising the bank as a "Modern Classic" called the A. T. Stewart house a "monument of the architectural uncultivation of the most conspicuous New York millionaire of A.D. 1870."[32]

But was Stewart the most conspicuous millionaire of the time, and was his house intended as a monument? The question is as problematic as Stewart's personal life is difficult to unravel. Stewart was a mystery to his contemporaries. After his death few could recall anecdotes or reminiscences that would illuminate his life and character. As is to be expected with a personality who was both so visible and so invisible, contradictory stories and rumors circulated. He was, by all accounts, a quiet, serious man, well read and an engaging conversationalist. He seems to have been largely devoid of personal eccentricities. An extensive description of Stewart appeared in the *Galaxy*. The reporter, who attributed Stewart's success to a "union of commercial genius and sterling character," suggested that for Americans, Stewart's was an ideal story: "As the career of this great merchant does not ask or suffer the 'heroic' style of treatment, so its value to young men is its simplicity, its lack of mystery, its story of penny upon penny and step upon step, and, finally, its possibility of imitation and the applicability of its moral to every-day life." In business Stewart was a difficult taskmaster who made few concessions to public criticism of his treatment of his employees, yet he seems by all contemporary testimony to have been forthright and honest. As the *New York Times* put it, "To his subordinates he was just, but not generous" (Fig. 24). Apparently he avoided using his wealth for political manipulation, although he did become an intimate of the Tweed ring. He was not shy, however, when it came to seeking legislation that would benefit his business and is credited with having stopped a planned rapid transit railway on Broadway. Stewart was appointed by President Grant to the position of Secretary of the Treasury, but he was barred from assuming office by a section of Hamilton's revenue act that disqualified persons whose income derived from importation. Stewart

[32] Van Rensselaer, "Recent Architecture in America, V, City Dwellings," *Century Illustrated* 31, no. 4 (Feb. 1886): 551–53; Schuyler, "Modern Classic," p. 589. See "Big Bonanza Buildings," *Real Estate Record* 17, no. 421 (Apr. 8, 1876): 255; Henry Watterson, *History of the Manhattan Club, a Narrative of the Activities of Half a Century* (New York: privately printed, 1915). See also the discussion of the Stewart house in Harry W. Desmond and Herbert Croly, *Stately Homes in America from Colonial Times to the Present Day* (New York: D. Appleton & Co., 1903), pp. 252–55.

SLAVERY IN THE PRESENT DAY; OR
"THE MODERN UN-PHARAOH."

SLAVERY IN THE OLDEN TIMES; OR,
THE ANCIENT PHARAOH.

FIG. 24. *The Modern Un-Pharaoh,* cartoon of A. T. Stewart. From *The Arcadian* (1870–73), loose sheet, n.d. (Print Department, New York Public Library.)

offered to place all of his business interests in a trust during his term and to devote the income to charitable and municipal purposes, but the offer was rejected. This was the closest he ever came to holding political office.[33]

Stewart's proposition was typical of the erratic benevolence to which he sometimes committed himself. Dramatic gestures, such as purchasing a ship, stocking it with provisions for victims of the Irish famines, and promising free return passage and employment, were repeated throughout his life. Sufferers of the Franco-Prussian War and victims of fires in Chicago and Boston were among those who had reason to be grateful to the mer-

chant prince, but no sustained philanthropic effort can be documented. The *New York Times* recalled: "He was ready at all times to help the public if at the same time he helped himself. He would foster American industry when there was a profit in doing it, and at one time, if not always, took all the silks which could be made at the famous mills of the Cheney Brothers at South Manchester, Conn." Stewart was active in the Union cause and gave liberally to the fund that New Yorkers collected for Grant after the war, but his prime charitable beneficiaries were workingmen and workingwomen. As early as 1846 he set up a boardinghouse adjacent to the marble palace; he also provided his clerks with a library. At one time he contemplated the construction of low-rent housing duplicating George Peabody's successful project in London,

[33] Crapsey, "Monument of Trade," p. 101; *New York Times,* Apr. 11, 1876; "Stewart as a Real Estate Operator," p. 237; Resseguie, "Decline and Fall," pp. 257–58.

FIG. 25. John Kellum, A. T. Stewart's Workingwomen's Hotel. New York City, 1869–73. From *Harper's Weekly* 22, no. 1111 (Apr. 13, 1878): 296.

but, having abandoned this scheme, he conceived the idea of a hotel for workingwomen. A building was constructed from designs by John Kellum (a cast iron hotel for women of cast iron reputation, according to one modern critic), but the building remained incomplete at Stewart's death and was not devoted to the idealistic cause he had envisioned (Fig. 25). Similarly, Garden City never functioned as the model workingman's community that Stewart had had in mind. Most contemporaries fully expected Stewart to make a major charitable gift or legacy. Peter Cooper, himself a great New York benefactor, commented optimistically: "I have long hoped and believed that Mr. Stewart had noble purposes formed which he intended to carry out, and which his great wealth and general knowledge of business had so admirably fitted him to accomplish." An account of Stewart in the *Cincinnati Gazette* put it more bluntly: "There are two events which will yet create a sensation here—

one will be the death of A. T. Stewart, the next the reading of his will." [34]

On April 15, four days after Stewart's death, his will was published in the New York papers. Murmurs of astonishment must have been uttered when the contents of the will were disclosed. Amid estimates of Stewart's fortune ranging from $25 million to $40 million, it was found that with the exception of minor gifts to friends and employees of long service, amounting to a little more than $300,000, and a bequest of $1 million to Judge Henry Hilton, his personal attorney and trustee, he left the entirety of his estate to his wife. Stewart endowed no charities and made no large benefactions, although he instructed his wife by letter to do so. A. T. Stewart, according to the *New York*

[34] *New York Times,* Apr. 11, 1876; Resseguie, "Decline and Fall," pp. 259, 272–76; Logan, "That Was New York," p. 82; "Sketch of the Greatest Business Man of America," p. 285.

*Herald*, "appears in the light of one who hesitated till the opportunity was gone. . . . He dreamed over magnificent schemes for the benefit of the city—public charities doubtless of a very practical nature—but he could not decide, and life passed away ere he could determine how to act, or to decide which of the many schemes was most to his satisfaction." The *Saturday Review* in London, commenting on the general vulgarity of American funerals as exemplified by reports of Stewart's ceremonial departure, offered a different explanation. "A great fortune is great only when intact; when spent it changes its character, for it is tied down to particular uses. It remains to be seen whether Mrs. Stewart and her trustee Judge Hilton, will have the courage to disperse the great accumulation which has been left at their disposal." [35]

By the time the *Saturday Review* column appeared, Judge Hilton had acquired the entire assets of A. T. Stewart and Company from Stewart's widow for $1 million, the sum of his inheritance from Stewart. Ignoring the provision in the will ordering liquidation of the business, Hilton tried to continue operating the Stewart interests. As an ironic testament to Stewart's business acumen and Hilton's ineptitude, within six years, the business was liquidated as a result of mismanagement. Mrs. Stewart, far from destitute, continued to live comfortably in the palazzo and to support the erection of a great Episcopal cathedral and cathedral schools in Garden City, where, she presumed, her husband's remains had been reinterred. (Stewart's body had been stolen from its crypt in Saint Mark's burying ground in New York in 1878. Despite rumors of ransom demands, the macabre act seems to have been one of vengeance undertaken by two Irish boys whose father's grave had disappeared in a cave-in during the construction of Stewart's cast iron store. In all probability Stewart's body was never returned. Mrs. Stewart, shocked by the theft of the body, was convinced by Hilton of its return.) She spent the remainder of her life warding off suits lodged against the estate by people hoping to prove kinship and claim a share of the once great fortune. [36]

Stewart's major act of philanthropy was, in certain respects, the construction of his house. But was it a public act of philanthropy or an attempt to disguise the workaday roots of Stewart's fortune? Was Stewart advertising his own success, and thereby his business, while gaining a bit of immortality for himself? What function did the house serve for a man who was childless and in his sixties when construction began? If the reminiscences of an old man can be trusted, Richard Lathers's suggestion that Stewart give the house to the city as a mayoral mansion had been rebuffed with the reply that Stewart had not slaved in business to decorate the municipality. On the other hand, Stewart made no attempt to cast off his image as a merchant or to dissociate himself from his business, which was, in fact, self-advertising. One critic had remarked: "Mr. Stewart, indeed, cannot be 'advertised,' in the ordinary meaning of that word. You might as well advertise the city of New York. While courting the widest publicity for his wares, he abhors personal publicity." Stewart would undoubtedly have been aghast at the cartoon in *Puck* implying that the saga of the grave-robbing was *The Story of a Great Advertising Dodge* (Fig. 26). It was largely Stewart's phenomenal success that made him visible to the public, yet he underscored that visibility through the construction of monumental business buildings. And, in the context of New York at the time, his house could not have been more conspicuous. Indeed, if he had needed the house as an entrée into society, nothing could have worked more toward the opposite end. The *New York Herald* explored these themes when it noted in Stewart's obituary:

As far as we can gather, Mr. Stewart seems to have been a very respectable man, who made his money by his

---

[35] *New York Herald* quoted in "A Millionaire's Funeral," *Saturday Review of Politics, Literature, Science and Art* (London), 41, no. 1070 (Apr. 29, 1876): 551. See also "The Estate of A. T. Stewart," *Real Estate Record* 17, no. 423 (Apr. 22, 1876): 299.

[36] Resseguie, "Decline and Fall," pp. 264–66. *Posthumous Relatives of the Late Alex. T. Stewart* contains most of the

legal documents. Resseguie effectively summarizes the evidence against Hilton and also discusses how the project for the chapel at Garden City was expanded to the proportions of a cathedral. For contemporary discussions of the cathedral, see "The Stewart Memorial Chapel, Garden City, Long Island," *Art Journal* 4, no. 2 (Feb. 1878): 47–50; "A Millionaire's Cathedral City," *American Architect and Building News* 6, no. 195 (Sept. 20, 1879): 102–3; Ruth Reynolds, "The Merchant Prince and the Grave Robbers," *Sunday News* (New York), Aug. 12, 1956, pp. 78–79; the *Morning Journal* reprinted the column "Inside the Stewart House" on October 26, 1886, after the death of Mrs. Stewart, and gave an extensive description of her life and activities after Stewart's death. See also obituaries in the *New York Times*, Oct. 26, 1886; *Harper's Bazaar* 19, no. 47 (Nov. 20, 1886): 761, 763.

FIG. 26. *A Great Advertising Dodge*. From *Puck* 5, no. 129 (Aug. 27, 1879): 393. *Puck* ran a series of cartoons in 1878–79 dealing with the Stewart grave robbery and Hilton's mismanagement of Stewart's business interests.

steady business habits and devotion to his shop. We can readily believe that in so far as greatness can be associated with a dry goods store, Mr. Stewart was a great man of his kind. But it is evident that, in the popular view of the matter his greatness consisted in his fortune rather than in his personal character; and in all probability, if he had been less opulent, he would never have been heard of beyond the range of his customers.

Yet how opulent was he? The house was marble and grand in scale, and its interiors were decorated to conform to that grandeur. Stewart's life-style, however, could not really be deemed opulent. According to the *New York Times*:

His mode of life has been simple and of an abstinent character. In his dress Mr. Stewart was simple in the extreme, but always neat. Almost his only personal luxury outside his home was the fine horses he used in driving to his stores, and he had several of these which were not surpassed in the City, although he never made any pretensions to or seemed to want speed.

Stewart reputedly spent his leisure time reading Latin and Greek; when asked what he would do if

he retired from business, he replied that he would go back to school.[37]

Although Stewart is said to have remarked that the house was "a little attention to Mrs. Stewart," it is likely that the motivation went much deeper. Comfort and elegance were certainly a consideration, but they were achieved with an exaggerated amplitude. In all of Stewart's endeavors, amplitude seems to have been a major force. He was enamored of big schemes and projects, such as the Garden

[37] Sanborn, *Reminiscences of Richard Lathers*, p. 46; Crapsey, "Monument of Trade," p. 101; *New York Herald* quoted in "Millionaire's Funeral," p. 550; *New York Times*, Apr. 11, 1876. "Alexander T. Stewart," *Harpers' Monthly*, p. 524, mentions Stewart's custom of reading Latin and Greek in leisure hours, a fact often repeated in Stewart biographies. While Stewart taught the classics in his early years in New York, the list of his library in the sale catalog includes only a handful of the most common classic works—many in translation. There is, of course, no guarantee that the entire contents were put up for sale or that in the ten years between the death of Stewart and that of his widow the books were not given away or dispersed.

City venture and the women's hotel. He greatly enlarged the Grand Union Hotel in Saratoga after he bought it in 1872 for $532,000 in cash. A profound believer in tangible assets, Stewart did not maintain a personal bank account, nor did he speculate in stocks. He bought real estate extensively and was condemned for his unprofitable investments, but it is likely that real estate was a way of solidifying surplus cash. There was not a cent in mortgages on the $10 million worth of New York real estate he owned, and he paid cash for everything he bought, including all of the merchandise in his store.[38]

A. T. Stewart seems to have derived his energy from constant activity. At age sixty-eight he was described as "constantly engaging in new enterprises that will require years for completion, and . . . steadily extending those all absorbing tentacles further and further. It is evident that he sees no shadow of death or retrogression lying athwart the near future." Stewart was a superstitious man, and his materialism may have had something to do with an emotional need to maintain personal control over his interests. His failure to prescribe specific philanthropic enterprises in his will suggests an unwillingness to speculate about the future or to deal with things beyond his personal control. In-

stead, his constant activity and planning became an emotional hedge against fate.[39]

Lacking personal papers that might yield a deeper insight into Stewart's motives, we must rely on incomplete contemporary testimony. The simple questions of why Stewart slaved in business to build a mercantile empire and why he waited for nearly thirty years after he became a millionaire to construct a great palazzo remain, for the time, unanswerable. Whether these were undertaken for the sake of conspicuousness is a moot point and one that should not concern us here, for the notion of conspicuous consumption has veiled much of the significance of the development of the millionaire's mansion in America. Montgomery Schuyler used the term pejoratively to attack the Stewart house while defending the Vanderbilt houses, which were subsequently condemned under the same indictment. A study of the Stewart house reveals that whatever Stewart's personal motives, there was a strong critical sentiment in favor of such mansions; far from being the incarnation of a new spirit of post–Civil War America, they were deeply rooted in American thought. Only by brushing aside simplistic critical terms can we begin to untangle the complex web of social, personal, political, economic, and emotional factors that contributed to the evolving fabric of urban domestic architecture in the second half of the nineteenth century in America.

[38] Logan, "That Was New York," p. 104; "Sketch of the Greatest Business Man of America," p. 165; "Stewart as a Real Estate Operator," pp. 278–79. The total value of Stewart's taxable property in New York was $15 million. The tax on Stewart's real estate in the year preceding his death was $180,392; on his personal property, $88,200.

[39] Crapsey, "Monument of Trade," p. 101; "Estate of A. T. Stewart," p. 299.

# La Farge's Eclectic Idealism in Three New York City Churches

*Helene Barbara Weinberg*

THE ECONOMIC, POLITICAL, and philosophical upheaval accompanying the Civil War conditioned the course of late nineteenth-century American art. The victory of the North in the war, a victory for the forces of industry and finance, was followed by other victories for machines and commerce. Railroad networks conquered distance; factory production vanquished handicraft; a materialist philosophy triumphed over agrarian values. Before the war, the need for a definition of what in the American experience was worthy of artistic interpretation had been satisfied by a highly naturalistic and moralistic interest in landscape. As the machine encroached on the garden, and as the Emersonian faith in the identity of the garden with national ideals faded, the reality of the garden became a less reassuring source of artistic content. As the native roots that had nourished American painting in the half century before the war atrophied, a search for new roots, new sources of inspiration, began. Those who had won the battles against secession and against the wilderness stimulated and directed this search. The material power of these industrialists and financiers demanded expression in cultural terms. They developed a taste for European art and imported it in increasing quantities to provide a stamp of sophistication and respectability for themselves.

Motivated by the growth of new national values and by their patrons' affection for European art, American artists increasingly explored foreign precedents for an art related at once to tradition and to contemporary American needs. A retreat from realism accompanied a heightened interest in ideal art, which was an expression more of the conceptual than of the purely perceptual, and which relied more on thoughts than on objects unmodified by the artist's ideational activity. John

La Farge (1835–1910) was among the most successful practitioners of this mode of eclectic idealism. His decorative works—thirteen major mural projects and numerous stained glass windows—reflect not only a remarkable artistic personality but the practical, aesthetic, and philosophical attitudes of an era of American culture. The number, scale, and location of La Farge's commissions testify to the growth of urban populations, the extension of national territory, and the creation of new civic and religious facilities. These developments sparked an enthusiasm for monumental art, and the taste that guided the decoration of elaborate homes expanded to encompass libraries, state houses, and churches. La Farge brought to his commissions a unique ability to allude to world history, religion, and mythology, yet he retained a respect for realist depiction. The results of the new taste suggest that the late nineteenth century in general and La Farge in particular produced a new fusion of the streams of ideal and real that had flowed concurrently through American art. His projects for secular decoration show that his patrons perceived and enjoyed parallels between their economic and cultural achievements and those of Renaissance merchant princes, correspondences between their clubs and the English club tradition, and affinities between American governmental institutions and the European political and legal heritage. His ecclesiastical decoration reveals liturgical changes and growing dissatisfaction with the traditional austerity of American churches, particularly among Episcopal congregations. An analysis of three such projects, New York churches that La Farge decorated between 1877 and 1888, is the focus of this paper.

La Farge's qualifications for responding to the complex and cosmopolitan attitudes of his time were rooted in the circumstances of his birth in New York City to well-to-do French émigré parents and of his childhood amid surroundings that encouraged interest in European literature, attention to religious practice, and appreciation of art.[1] By 1859, when he decided to become a painter, he had received a liberal college education, had studied law, and had visited Paris. There, besides working briefly in the studio of Thomas Couture and copying in the Louvre, he had been introduced to the leading literary and artistic personalities and issues of the day. Plunged into the continuing controversy between neoclassic and romantic artists, he managed to retain respect for the former while allying himself with the spirit of the latter, particularly Delacroix, "whose astonishing paintings," he recalled, "had been, with those of the old masters, one of the first great sensations of my first days in Paris." [2] La Farge's receptivity to a variety of styles was matched by his curiosity regarding the media of their expression. His examination of medieval stained glass and his investigation of mural painting techniques during his stay abroad significantly affected his development as a decorative artist.

La Farge's landscape and still life paintings of the 1860s, particularly those done in Newport under the guidance of William Morris Hunt, reflected his concern with mastering the principles and practice of realist painting. It was during this period too that La Farge developed a profound affection for Japanese art, particularly for prints, which, he found, reinforced his experiments in realism. Japanese art also appears to have quickened La Farge's interest in a more ideal and decorative mode of painting. This interest of his was further excited by an intensive reexamination of his personal religious belief and by a number of commissions for altar paintings and book and magazine illustrations.

By the early 1870s, La Farge had also experimented with mural decoration in several domestic settings. More important, he had established contact with H. H. Richardson, who shared his willingness to look to the styles and methods of production of traditional art to express the spirit and needs of the present. Richardson provided La Farge with his first commission for large-scale mural work, the decoration of the interior of Trinity Church, Boston. Here, in an incredibly short period—from midsummer 1876 to February 1877—the artist conceived, elaborated, and essentially concluded a unified decorative scheme of immense and unprecedented scale, suitable to the architecture of the church, to the theological attitudes of its Episcopal congregation, and to the abilities of his inexperienced workmen. Upon consecration of the church, critical response combined great admiration of the artistic effect of his work with an awareness of its novelty in the history of American art and its place as a forerunner of a new movement in religious decoration.[3]

In the decade following completion of the main part of his work at Trinity, La Farge was invited to provide mural paintings for three neo-Gothic Episcopal churches in New York. These projects presented a number of analogous problems and evoked solutions embodying his mature attitudes toward religious decoration. La Farge was stimulated to explore and recombine a variety of precedents so that his work might provide a necessary spiritual link with the past and yet remain appropriate to the architecture of each church. The projects challenged him to express ideal themes in convincing terms based on accurate depiction of realistic light and suggestive atmosphere, interests rooted in his early experiments in landscape painting. They stimulated the development of his personal mural style, based on the painterly colorism of the sixteenth-century Venetians and Delacroix. They presented opportunities to declare allegiance to tradition not only in form and style but in methods, and to demonstrate practically his theoretical commitment to the collaboration of artists in various media and to the coordination of the roles of artist and artisan for a unified effect.

---

[1] For a summary of La Farge's early development, see Helene Barbara Weinberg, "John La Farge: The Relation of His Illustrations to His Ideal Art," *American Art Journal* 5, no. 1 (May 1973): 54–73.

[2] La Farge's autobiographical memoranda, Royal Cortissoz Correspondence, Beinecke Rare Book and Manuscript Library, Yale University (hereafter Cortissoz Correspondence). This and all subsequent excerpts from this archive are quoted from the originals rather than from the transcriptions in Royal Cortissoz, *John La Farge, A Memoir and a Study* (Boston: Houghton Mifflin Co., 1911).

[3] See, for example, *Boston Evening Transcript,* Feb. 6, 1877; *Boston Daily Globe,* Feb. 10, 1877; Clarence Cook, "Recent Church Decoration," *Scribner's Monthly* 15, no. 4 (Feb. 1878): 570; George Parsons Lathrop, "John La Farge," *Scribner's Monthly* 21, no. 4 (Feb. 1881): 514. For a consideration of this project, see Helene Barbara Weinberg, "John La Farge and the Decoration of Trinity Church, Boston," *Journal of the Society of Architectural Historians,* in press.

Saint Thomas Church, at the northwest corner of Fifth Avenue and Fifty-third Street, was the last important church designed by Richard Upjohn. Although the church's first service was held in October 1870, its interior decoration remained incomplete. Before the spring of 1877, one of the parishioners, Charles H. Housman, provided a bequest for the decoration of the chancel as a memorial to his mother.[4] The precise date of initiation of the project is unknown. An early biographer's statement that La Farge received the commission when "his work at Trinity was but just finished" refers only to the main part of the Boston project, concluded in February 1877.[5] Two nave murals for Trinity Church, depicting the encounters of Christ with Nicodemus and the woman of Samaria, were painted concurrently with the chancel paintings for Saint Thomas. The first of the two murals in each church was completed in October 1877; neither of the companion paintings was finished before February 1878.

The success of his well publicized work in Boston was instrumental in bringing La Farge the New York commission. Contemporary critics often compared the two projects, noting that in Saint Thomas, "Mr. La Farge found himself working under conditions very different from those he met with in Trinity, and far less fortunate." Upjohn's retirement before the initiation of the chancel project precluded collaboration between architect and artist in the interest of a unified effect. The completion of the general interior scheme before La Farge was engaged to decorate the chancel and the contractual limitation of his work to the chancel area were further impediments to unity. In some quarters the limitation to the chancel was not considered an absolute detriment. One critic noted,

"It is perhaps quite as well that he was restricted to the chancel here; decoration of the entire church would have been an extremely difficult task—a task which admitted of hardly more than the rather unsatisfactory accomplishment of obliterating and softening bad effects."[6]

These "bad effects" in the church derived from a "construction of the thinnest, most pasteboardy kind" and a plan "inimical to the arts, which find no hospitality in these cross-lights, these cramped corners, these elbowing angles, and irreconcilable windows." The design of the chancel itself presented difficulties. A contemporary observer noted:

The apse of the church is polygonal, and has five sides; it is stone to the roof, plastered within and ceiled with wood. Twenty feet from the floor perhaps are five windows, with three lancets each, and above these are traced with wooden strips apparent relieving arches, from whose points also wooden strips imitate ribs of stone vaulting and meet overhead. The point of meeting . . . is considerably above the point of the arch which separates the chancel from the nave, so that it and much of the chancel ceiling is hidden from the view of any one not within the rail. The chancel being deep and the different panels few and wide, the two extreme panels make a large surface that is either invisible or foreshortened to the observer at the further end of the nave. The light is very high and confused; it comes in from the semi-circle—let us say—of windows and is opposed by a higher still row of gas-jets which follows the line of the arch that bounds the chancel and is invisible to the congregation.[7]

Handicapped by the physical attributes of the space to be decorated, La Farge was also restricted in the matter of expense. He found the sum of $3,500 insufficient for execution of his scheme and suffered personal financial loss. As he told a correspondent, "The donor came abruptly to the end of his money, and in fact I never was paid either adequately or in full for what I spent. St. Gaudens was paid but not largely. It was a sort of labor of love for us and we were very glad to put in our time and money to help in the new movement of decorating religious buildings in an artistic way."[8] As at Trin-

[4] See Everard M. Upjohn, *Richard Upjohn, Architect and Churchman* (New York: Columbia University Press, 1939), pp. 178, 180; William Walton, "The Decorations of the Chancel of St. Thomas' Church, New York City: Work of John La Farge and Augustus Saint-Gaudens," *Craftsman* 9 (Oct. 1905–Mar. 1906): 373.

[5] Cecilia Waern, "John La Farge, Artist and Writer," *Portfolio*, no. 26 (Apr. 1896): 37. A dearth of archival material precludes definition of many details of the project. The fire that destroyed the church in 1905 may have destroyed relevant old records. Surviving vestry minutes indicate little involvement on the part of that body in the decoration. The personal files of the rector and the donor, which might include pertinent correspondence and statements, have not been located. Robert C. Jones, sexton-administrator of the church, to author, Aug. 17, 1970.

[6] Cook, "Recent Church Decoration," p. 574; *New York World*, Oct. 28, 1877. Charles Henry Caffin's assertion of architect-artist collaboration, in *The Story of American Painting* (New York: Frederick A. Stokes Co., 1907), p. 308, was refuted in E. M. Upjohn, *Richard Upjohn*, p. 180.

[7] Cook, "Recent Church Decoration," p. 574; *New York World*, Oct. 28, 1877.

[8] *New York World*, Oct. 28, 1877; La Farge to Alfred Roelker, Aug. 12, 1905, La Farge Family Papers, Yale Uni-

ity, he was allowed only a short time to complete the work—"a few weeks for everything." In addition, he was in ill health[9] and experienced problems of communication with assistants and collaborators.

Requirements imposed by the rector of the church, William F. Morgan, determined the artist's plan for decorating the chancel. La Farge noted that the subject of the Resurrection "was chosen by the Rector . . . and he also chose the texts that give the motive of the Garden, in which I . . . placed the meeting of Mary with our Lord, and . . . directly connected it with the Tomb and the angels seated upon it and the sleeping guards." Morgan also requested the inclusion of a tall Latin cross, "because his parishioners were accustomed to it," and a Gothic bishop's chair, "the chair taking the place of the altar and an altar table being placed in the chancel. . . . The whole of my design," La Farge recalled, "was, therefore, built on these two points, to make the cross and the Bishop's chair of extreme importance." Restrained by specific physical and iconographical limitations, La Farge exploited the pentagonal form of the apse for an effect reminiscent of an early Renaissance altarpiece (Fig. 1). In the central section, above the bishop's chair, he set the cross in a relief panel of adoring angels, "which was also a half-expressed wish on the part of Dr. Morgan," [10] and enframed the panel with carved and inlaid pilasters. At Morgan's request, again, he hung a heavy, ornate crown above the cross. The rector, La Farge noted, "wished a crown—to recall the words, no cross no crown, and he also wished something to break in his parishioners to a more generous and 'higher' Church view of the interior of a Church." [11] The wings of the altarpiece illustrated the texts selected by the rector, the *Noli Me Tangere* (John 20 : 17) in two panels at the left

(Fig. 2) and the *Visit of the Three Maries to the Tomb* (Luke 24 : 1–4) at the right (Fig. 3).

La Farge recalled that he was at first inclined to place these scenes in reverse order: "One was to see first the division where the Holy Women are warned of what has happened, and then of this further development of what is happening; the angel seated at the Tomb having been seen by Mary Magdalen, she turns away and sees our Lord who moves out of the picture as in the text. For that purpose I had made my design in sequence." [12] This arrangement would have produced a progression, from left to right, of the three Marys, the "two men . . . in shining garments" (Luke 24 : 4), the angel on the tomb, and the *Noli Me Tangere*. For visual continuity the tomb was placed toward the center of the design. It remained there in an alternative solution, recorded in a signed drawing dated May 4, 1877 (Fig. 4), in which the *Noli Me Tangere* was moved to the left side and the *Three Maries* scene reversed, so that the "two men" were placed toward the center and their gesture in the direction of the tomb was logical.

Despite the narrative sense of placing "the figure of our Lord . . . on the outside just before his disappearance from Mary's eyes," and despite the visual embodiment of spiritual belief in an arrangement of light and shadow giving "the figure of our Lord a more evanescent look as he came into the penumbra on the edge of the painting," neither of these solutions was acceptable to Dr. Morgan. La Farge wrote to Russell Sturgis: "Dr. Morgan was dissatisfied at the arrangement [of the two scenes], which chronologically & dramatically more correct, did not place the figure of Xt in the centre of the Chancel, the place of Honour, hence he requested to have them changed." [13] The resulting reversal of the left-hand scene, which placed Christ close to the center, produced the least coherent narrative arrangement and interfered with the visual unity of the two scenes. Despite the continuity of the landscape setting, each scene required a separate viewing. La Farge was less disturbed by these changes than one might expect: "I do not think it has mattered much," he said, "but I feel that it would have been much better the other way, and so much more within the meaning of the text and

versity Library (hereafter LFP). See also La Farge, "Memorandum about St. Thomas' Chancel and St. Gaudens' Bas-Relief," Oct. 10, 1907, August F. Jacacci Papers, Archives of American Art.

[9] La Farge to William B. Van Ingen, quoted by Van Ingen in an article prepared for *Burlington Magazine* in 1905 but not published (see p. 2 of Van Ingen's typescript). Copies of the typescript, "The Destruction of a Masterpiece of American Art," are in August F. Jacacci Papers, Archives of American Art, and in LFP. See also La Farge to Russell Sturgis, Aug. 23, 1905, La Farge to Roelker, Aug. 12, 1905, LFP.

[10] La Farge quoted in Van Ingen, "Destruction of a Masterpiece," p. 5.

[11] La Farge, "Memorandum."

[12] La Farge quoted in Van Ingen, "Destruction of a Masterpiece," p. 3.

[13] Quoted in Van Ingen, "Destruction of a Masterpiece," p. 3; La Farge to Sturgis, Jan. 19, [1899], LFP.

FIG. 1. Saint Thomas chancel, New York City, before 1896. (Photo, courtesy Everard M. Upjohn.)

FIG. 2. John La Farge, *Noli Me Tangere*. Saint Thomas chancel, New York City, 1877. Probably encaustic on canvas; dimensions unknown. From *New England Magazine,* n.s. 12, no. 2 (Apr. 1895): 137.

FIG. 3. John La Farge, *Visit of the Three Maries to the Tomb*. Saint Thomas chancel, New York City, 1877–78. Probably encaustic on canvas; dimensions unknown. From *Bookman* 28, no. 2 (Oct. 1908): 129.

FIG. 4. John La Farge, study "for St. Thomas," 1877. Charcoal; H. 6″, W. 10⅞″. (Collection of Avery Library, Columbia University.)

the meaning of the picture that I think it is worth mentioning." [14]

If the general scheme was reminiscent of an altarpiece of the early Renaissance, particular elements also recalled that period. In composition and in details such as the hilly backdrop, the rectilinear sarcophagus with the angel on its edge, and the frontal Christ, the *Noli Me Tangere* echoed Giotto's Arena Chapel depiction of the same scene (Fig. 5). In his design for the relief panel of angels, whose execution he assigned to Augustus Saint-Gaudens, La Farge expressed a preference for the early Renaissance style (Fig. 6). On August 13, 1877, he advised the sculptor: "Do not take much stock in what Dr. Morgan thinks suitable for the figures unless you yourself approve of what he says. He has, as you remember, a fear that they will be too Catholic. There is no danger. There is no such thing as the Protestant in art. All you need do is not to make any aureoles around their heads. Any mediaeval sculpture, or, renaissance (not a late one), or paintings of the early time (Italian), give the type that will be needed to be neither high nor low church." Saint-Gaudens wrote La Farge, in a letter of August 29, 1877, "I have been to the Louvre to see the pictures of the early Renaissance, and the next relief will be more in that character." He later regretted that he had not studied Italian

[14] La Farge quoted in Van Ingen, "Destruction of a Masterpiece," p. 4.

FIG. 6. Augustus Saint-Gaudens, *Adoration of the Cross by Angels*, 1877. Wood engraving by Timothy Cole. From Homer Saint-Gaudens, ed., *The Reminiscences of Augustus Saint-Gaudens*, 2 vols. (New York: Century Co., 1913), 1:197.

models before executing the angels. In January 1878 he noted his impression of the Pisa Cathedral in a letter to La Farge, written from Rome: "There are on the vault over the high altar at the entrance of the chancel, angels by Cimabue . . . that are very much in the character of what we have done in the sculpture . . . had I seen them before I did my work I would have avoided much that is bad. . . . I wish I had to do the angels over again and could do them here." [15]

[15] Homer Saint-Gaudens, ed., *The Reminiscences of Augustus Saint-Gaudens*, 2 vols. (New York: Century Co., 1913), 1:190–93, 206, 209. La Farge's early admiration for the Padua frescoes and for "the directness of the earliest of masters" is recorded in Waern, "La Farge," pp. 13–14. The angels in the apse of Pisa Cathedral have since been attributed to Cimabue's associates.

FIG. 5. Giotto, *Noli Me Tangere*. Arena Chapel, Padua, ca. 1305. Fresco; H. 80″, W. 91″. (Collection of Arena Chapel, Padua.)

The early Renaissance style, evocative of a period of simple and direct belief and worship, seemed to La Farge an acceptable prototype for work in an American Protestant church, whose neo-Gothic design and theological inclinations it did not contradict. His scheme for Saint Thomas's chancel represented one of the earliest, if not the first, adaptation of the early Renaissance style for church decoration in America. Soon after the church's completion, it was lauded as an effort "to treat church decoration in the purest style of Ecclesiastical Art." Some years later it was hailed as "one of the very first [church designs] to be conceived and carried out in what was known as the 'new style.' " [16]

The concept of a new style implied more than the adaptation of Renaissance motifs to modern decorative needs. It embodied the larger Renaissance attitude toward the painter as a workman in the arts, a planner of unified schemes, and a collaborator with workmen in other media. La Farge devoted as much attention to his plan for the chancel as a whole, and to his designs for the enframement of his paintings, as to the paintings themselves. The artist testified to his participation in the execution of decorative details: "I was so much interested in it that a good deal of the carving, etc., was done by my own hand. I could not find anyone to give exactly the character I wanted to the work, which was of a manner and kind not known to the artists or the architects." [17] The most ambitious of the architectural moldings designed and largely executed by La Farge was the pair of pilasters enframing the central panel of angels (Figs. 2, 7). Carved in a scroll pattern, they were "superbly colored and completed by means of iridescent pearly shell, let into the wood in bits." In addition, he planned a "crowning cornice" and a set of pillars "bracketed out from the wall" to frame the reredos and the bishop's chair; for lack of funds, these were never executed. [18]

La Farge's concern with light transmitted

[16] A[nna] Bowman Dodd, "John La Farge," *Art Journal*, n.s. 9 (Sept. 1885): 263; Walton, "Decorations of the Chancel," p. 373.

[17] Lathrop, "La Farge," p. 515; La Farge quoted in Waern, "La Farge," p. 38. See also Pauline King, *American Mural Painting* (Boston: Noyes, Platt & Co., 1902), pp. 30–31; La Farge to Roelker, Aug. 12, 1905, LFP.

[18] Lathrop, "La Farge," p. 514; Waern, "La Farge," p. 39. *New York World*, Oct. 28, 1877, describes the pilasters as "gold and dark."

FIG. 7. John La Farge, section of pilaster shown in figure 2. Saint Thomas chancel, New York City, 1877. From *Scribner's Monthly* 15, no. 4 (Feb. 1878): 573.

through stained glass had its roots in his work at Trinity Church. There he had participated in designing glass to complement his color scheme for the tower walls and, later, had proposed grisaille glass to reconcile the color of nave windows and walls. For Saint Thomas, limited by the presence of stained glass already in place, he produced a successful, if makeshift, remedy. He obtained permission to paint over the windows in the chancel in order to integrate them with his decorative scheme. While retaining their linear design, he tempered light, neutralized tints, and altered obtrusive details, creating a "noble, deep, rich, mellow" impression.[19]

To further harmonize the colors of windows and murals, La Farge planned painted decoration for the chancel ceiling, and when the "French decorative painter" to whom he had assigned the execution of his designs misinterpreted his drawings and color notes, La Farge repainted the area almost entirely by himself. One critic described the "tenderly glowing deep topaz" effect resulting from the application of deep blue foliate arabesques to vaults painted red and gold and separated by blue and white ribs, adding: "This ceiling, with its amazing soft reflections of color thrown back and forth by the inclinations of the vaulting . . . is in itself a noble decorative picture." [20] The artist shared this high estimate of the effectiveness of the ceiling: "I doubt if there is another merely ornamental ceiling in America painted by an artist of repute, with his own hand; and in fact there must be very few anywhere. I was always very proud of both the being able to do with without breaking down from hard work, and also of the rescuing of the work from the Frenchman's blunder, and then really of the result, which has always seemed to be very handsome —far handsomer than anything I know of, except a few very great examples." [21]

The collaboration between La Farge and Augustus Saint-Gaudens in the design and execution of the reredos is well documented. La Farge prepared a general plan, "merely a design—arrangement and number of figures and grouping"—incorporating the elements recommended by the rector. Limited funds dictated modeling the reliefs in plaster and coloring them to harmonize with the painted decorations. Saint-Gaudens received the commission shortly before he left for Europe in June 1877. By mid-July he had arrived in Paris, where, he recalled, he "took a little studio . . . , and began at once on the St. Thomas' reliefs." La Farge's advice and criticism, based on descriptions and photographs Saint-Gaudens sent from Paris, guided the sculptor. Although anxious to conform to La Farge's ideas and requirements, Saint-Gaudens was allowed some latitude in interpreting the design. By the end of August, Saint-Gaudens had completed the first panel. On La Farge's recommendation, he viewed the early Renaissance pictures in the Louvre and prepared the second relief in that style. In the interest of better proportion for his figures, he added ornamental panels at the outer edges of the bas relief; to obtain equal division of the reliefs and to emphasize horizontality, he added a frieze of cherubs' heads and wings above. He also engaged Will H. Low, an American artist completing his studies in Paris, to paint the reliefs.[22]

The second relief was finished by September 6, and Saint-Gaudens expressed satisfaction at the effect of the coloring: "If you are pleased, which I think you will be, they being a great deal better colored and having a great deal more character than as you see them in the photograph, I shall be happy. . . . I have tried to keep them so as not to clash with your painting. They have been painted at the height, or very nearly so, at which they will be seen; and must not be judged before they are in place, being very coarsely done when seen close to." On September 10, Saint-Gaudens wrote La Farge that despite the rapidity of execution—an average of two and a half days for modeling each of the large panels and five hours for the frieze of cherubs' heads—he was confident in the success of the work, but "anxious about the impression it will produce." Less than two months after he had begun work, Saint-Gaudens shipped the panels to New York. In an accompanying letter, dated September 20, he expressed concern over La Farge's reaction: "If it is not entirely as you wish, lay it to the tremendous haste in which it was done. I make no pre-

[19] Helene Barbara Weinberg, "The Early Stained Glass Work of John La Farge (1835–1910)," *Stained Glass* 67, no. 2 (Summer 1972): 7–10; *New York World*, Oct. 28, 1877.

[20] La Farge's description of the incident is quoted in Van Ingen, "Destruction of a Masterpiece," pp. 6–8; *New York World*, Oct. 28, 1877.

[21] La Farge quoted in Van Ingen, "Destruction of a Masterpiece," p. 8.

[22] La Farge, "Memorandum"; H. Saint-Gaudens, *Reminiscences of A. Saint-Gaudens*, 1:190–94.

tense of it as a piece of sculpture, that would be ridiculous considering the time I had to do it, but my aim has been to keep it grave, harmonious in tone, and above all good in general effect, having your work in my mind all the time." [23]

That the collaboration by mail had not produced the desired result is evident in La Farge's petulant reply, which mixed praise with considerable criticism. The general impression, he said, was consistent with his original intention. "The whole appearance is so successful that at a distance there is a breath of Italy in it which takes hold of every one." More specifically, he condemned the change in the proportions of the angels, which made them look small compared to the stained glass and the pictures, and the addition of the band of cherubs' heads separating "the work from the crown, which now has no support and seems to fall. All that I assure you was thought of by me when I made my drawing and was the pivot of my entire work." Difficulties also arose in accommodating the ornamental moldings to the reliefs: "I had to displace them and cut them and put them by trials to the wall until they composed tolerably. . . . I did all I could and three weeks of work were spent; it cost me nearly what I paid you to make all this alteration." Most irritating to La Farge was the fact that he had to cut one of his pictures: "As the principal figure came within two inches of the edge, [it] was very disagreeable to my artistic feelings and to my personal feeling also. As I had given you the principal place, it was a bore also to have even my place injured." La Farge also altered the painting of the relief panels to diminish the "doorway look produced by all those squares with angels in them which makes it look as if the wall were a door and prevents solidity." Saint-Gaudens's answer suggested that his failure to please La Farge was the result of poor communication regarding each artist's intentions and plans. He welcomed La Farge's alterations and felt gratified that the work had at least been in general harmony with the painter's desires.[24]

The reredos was the first element of the decoration to be installed. Sculptured in high relief on either side of the cross were three pairs of kneeling angels, in free design and small scale. Single angels and cherubs' heads occupied the lower panels. In the band beneath the crown, the irregular distribution of cherubs' heads countered the symmetrical arrangement of the main portion of the work and lent a suggestion of movement. While one critic considered the pale, cool coloring of the crown, angels, and cherubs in "a delicate blue and gold" disappointing, a painter and friend of Saint-Gaudens's told the sculptor that he had "never seen any sculpture so well managed for similar circumstances of light and distance." [25]

Although the composition of the *Noli Me Tangere* was clearly based on precedent, the painting exhibited a quality peculiar to La Farge's mature mural works, a reminiscence of tapestry in which bits of color in the background and foreground assume a life of their own, modifying the realism of the scene by increasing decorative flatness. While premonitions of this effect had appeared in some of La Farge's easel paintings of the early 1870s—the *Virgil* (Fogg Art Museum) and the *Muse of Painting* (Metropolitan Museum of Art)—it reached its fullest expression in the murals in Saint Thomas. A diffused atmospheric glow, consistent with the early morning setting of the scene, was exploited to bathe and dissolve forms and to give an overall impression of scintillating light. The freshness of the colors was enhanced by the use of wax rather than oil as a medium. The want of "solidity in the painting," which was criticized by one writer,[26] appears to have been intentional. Compelled, in light of contemporary aesthetic demands, to abandon the abstract background of the Giottesque conception, La Farge was able to produce a landscape setting that gave a real sense of place without contradicting the flatness of the wall surface.

The *Noli Me Tangere* was installed on October 29, 1877; the painting of the *Three Maries* was not completed and placed until the spring of the following year.[27] La Farge found his source for the

[23] H. Saint-Gaudens, *Reminiscences of A. Saint-Gaudens,* 1:194–200. Low described his role in persuading Saint-Gaudens to treat the reliefs in polychrome rather than gilt, as well as their efforts to simulate the conditions under which the reredos would be seen, in *A Chronicle of Friendships, 1873–1900* (New York: Charles Scribner's Sons, 1908), pp. 222–23, 226.

[24] H. Saint-Gaudens, *Reminiscences of A. Saint Gaudens,* 1:201–3.

[25] *New York World,* Oct. 28, 1877; Wyatt Eaton to Saint-Gaudens, undated, in H. Saint-Gaudens, *Reminiscences of A. Saint Gaudens,* 1:209–10.

[26] The medium is recorded in "a newspaper article" quoted in Henry A. La Farge, "John La Farge: catalogue raisonné," LFP; Cook, "Recent Church Decoration," p. 574.

[27] *New York World,* Oct. 28, 1877. Cook, "Recent Church

FIG. 8. *The Marys at the Sepulcher,* Italy, ca. 400. Ivory; H. 7⅜", W. 4 9/16". (Collection of Bayerisches Nationalmuseum, Munich.)

FIG. 9. John La Farge, *Three Maries.* Watercolor; H. 9 3/16", W. 8 5/16". (Collection of Henry A. La Farge: Photo, Metropolitan Museum of Art.)

figures of the three women in an early Christian ivory panel (Fig. 8). To the general arrangement derived from this conception, he added graceful movement and rich color. The colors of the scene may be deduced from a watercolor study showing the figures in heavy draperies of blue violet, burgundy red, and moss green (Fig. 9), and from contemporary comments that the painting was "bolder

and more stirring" than the *Noli* panel, as well as brighter in light and more earthly in atmosphere.[28] La Farge's compromise between the supernatural implications of the narrative and the demands of realistic painting anticipated a similar compromise in his painting of the *Ascension* a decade later. It was recognized by critics who lauded the artist for "going beyond the real, yet arresting the real aspect, also, and fixing it in a dimly luminous beauty." [29]

As has been noted, La Farge devoted considerable attention to designs for the settings of these paintings. Years after the completion of the work he said he "always regretted that the funds of the donor to the Church had not allowed their being carried out as projected." It was surely with ambivalent feelings that he responded to the church's proposal in 1896 that he plan alterations for the chancel which he had never finished decorating. He was first consulted in May 1896 and, by early June, was "about to prepare the scheme." [30] The church's

Decoration," p. 574, noted in February 1878 that as yet only one picture was finished. On April 26, 1878, the vestry offered a resolution of thanks to the donor for the decoration of the chancel "now nearly complete." Quoted in Robert C. Jones to author, Aug. 27, 1970. La Farge had adopted the procedure of executing large-scale works on canvas for later application to the wall. This was the method of all his mural painting subsequent to Trinity, where paint was applied directly to the plaster. Only the *Ascension* is known to have been executed in place, albeit on canvas.

[28] H. La Farge, "John La Farge"; Lathrop, "La Farge," p. 514; King, *American Mural Painting,* p. 32.

[29] Lathrop, "La Farge," p. 514.

[30] La Farge to Roelker, Aug. 12, 1905, Bancel La Farge to J. Wesley Brown [Rector] (letterpress), May 20, 1896, Bancel

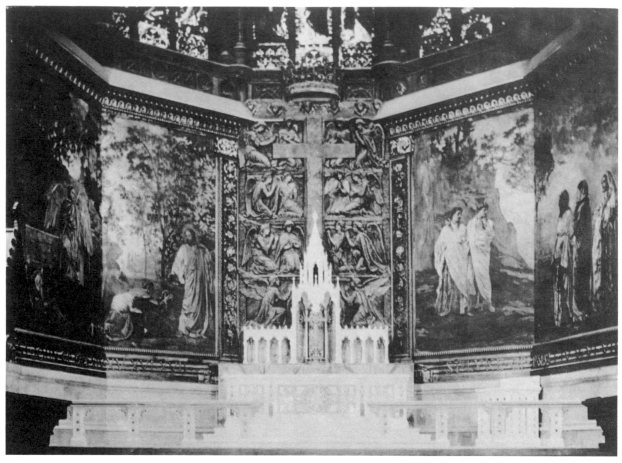

Fig. 10. Saint Thomas chancel, New York City, after 1896. (Photo, courtesy Everard M. Upjohn.)

projected changes included reproduction of the reredos in bronze, a proposal to which Saint-Gaudens objected, stating that gilded and painted marble would suggest the original effect more closely. In the final scheme the original reredos was never reproduced. Instead, the bishop's chair was moved to the right of the chancel and its place filled with a neo-Gothic marble and mosaic altar (Fig. 10). La Farge was interested in securing the commission to design this altar but was informed by July 10, 1896, that the contract had been given out. Although one critic stated that "the difficult task of harmonizing this addition to the old work was successfully carried out by Mr. Chas. R. Lamb, with the full approval of Messrs. Saint-Gaudens

and La Farge," [31] La Farge's comments betray little enthusiasm for Lamb's work. He referred to the altar as "beastly" and called the rector "an amiable ass" who "spoiled the whole thing by giving to those dreadful Lamb boys, an altar arrangement, placed as they alone could manage it—in dead white—of dismal shape style fashion & dimension—just where it should not be." Whatever plans La Farge devised for the alteration of the chancel were rejected, and his role in the redecoration episode was limited to the execution of a small circular window, for which his son, Bancel, assumed responsibility.[32]

La Farge to R. W. Gibson [Vestryman] (letterpress), ca. June 6, 1896, LFP. Bancel La Farge served as his father's chief studio assistant through the 1890s.

[31] Bancel La Farge to Gibson (letterpress), July 10, 1896, LFP; Walton, "Decorations of the Chancel," p. 373.

[32] La Farge, "Memorandum"; La Farge to Sturgis, Jan. 19, [1899]; Bancel La Farge to Brown (letterpress), Aug. 21, 24, Sept. 14, Oct. 1, 1896, LFP.

FIG. 11. St. Thomas chancel after fire. New York City, 1905. (Photo, courtesy Everard M. Upjohn.)

Apparently the church was not satisfied with the results of the 1896 redecoration, for in 1904 La Farge was again consulted regarding alterations. On August 2 he complained to a friend: "The people at St. Thomas have used up my time and energies, of which I have not much, and St. Gaudens' also, in making arrangements supposed to be in a frantic hurry for the decoration of the Church and the various changes and repairs and alterations and emendations, etc." In 1907 he recalled: "3 years ago the church thought of completing the work and St. Gaudens and I went to look at it. I proposed to him to remodel and [illegible] with an altar to be designed by us and carried out by him in detail—so that the whole affair might be art. Other details were also to be put into lasting mate-

rials. But that was my part." Other important elements of the project were La Farge's plans to rearrange the lighting and to remove combustible materials from the chancel. La Farge believed that if this renovation had taken place, the fire that destroyed the church on the morning of August 8, 1905, might not have occurred (Fig. 11).[33]

[33] La Farge to Mrs. Cadwalader Jones, Aug. 2, 1904, LFP; La Farge, "Memorandum." See La Farge to Sturgis, Aug. 23, 1905, LFP. The fire was believed to have been caused by defective electrical wiring behind the south half of the organ. Only the tower remained after the fire; from the interior, only the altar cross and a bronze bust of Dr. Morgan were saved. The loss was estimated at $400,000. See *New York Herald*, Aug. 9, 1905; Harold E. Grove, ed., *St. Thomas Church* (New York: by the church, 1965), p. 2.

La Farge's replies to expressions of sympathy after the fire reveal his estimate of the importance of the chancel decoration. He wrote to Henry Adams in 1906:

Summer before last, fire, a foolish sexton, and the still more foolish vestry, managed to burn up my work and Saint-Gaudens' at St. Thomas's Church. . . . I felt very badly because it seemed to me the only large piece of work—I mean painting—which I had a chance of doing, and which represented what I thought I could do in the art of painting, which is one of continuous development; and I had done something new which nobody else had done, and which I, to-day, would not feel bold enough to undertake. Nobody in the future will ever know what I have done.[34]

He supposed the church would call on him to restore the work, but had doubts regarding his, or any other artist's, ability to repeat or reproduce a painting in all its original spirit. "In this particular case it would be extremely difficult," he told Russell Sturgis, "because the pictures were painted with a good deal of 'go' and rapidity and under a certain excitement which gave them, I think, just that quality which Saint-Gaudens valued, apart from the question of design and the meaning and the arrangement of light and shadow and the, I think, successful representation of time of day intended."[35] La Farge was excluded from the project for rebuilding the church, which was undertaken between 1909 and 1914 by the architectural firm of Cram, Goodhue, and Ferguson.

The Saint Thomas murals were an unfortunate loss for students of La Farge's art and of American church decoration. In this first of several projects marking his maturity as a mural painter, La Farge demonstrated allegiance to tradition in form, style, and method, assuming responsibility not only for the design of a unified scheme but for its actual execution. He effectively integrated ideal, supernatural themes into settings notable for realistic light and suggestive atmosphere. He depicted time and place without contradicting the planar quality of mural work. While avoiding vapid historicism, he relied on a combination of precedents to provide a desired spiritual link with the past, amply fulfilling the rector's wish to bring his congregation to

a " 'higher' Church view" of decoration without compromising the principles of Protestant worship.

Seven years and two other church projects—the United Congregational Church, Newport (1880), and the Brick Presbyterian Church, New York City (1883)—intervened between the decoration of Saint Thomas and that of the Church of the Incarnation. As the United Congregational and Brick Presbyterian commissions involved enrichment of the interiors with abstract decorative patterns, they do not concern us here. In Saint Thomas and the Incarnation, however, La Farge confronted analogous problems of figurative mural painting and produced similar solutions. These two projects are, therefore, closely linked.

The Church of the Incarnation was constructed on its present site, at the northeast corner of Madison Avenue and Thirty-fifth Street, in 1863–64. Its interior, in the early decorated English Gothic style, relied on memorial windows set into stenciled walls for ornament. The shallow, windowless chancel was austerely decorated with Gothic medallions and moldings, and furnished only with a simple ceremonial table and carved pulpit (Fig. 12). On March 24, 1882, fire damaged the church, particularly in the area of the chancel.[36] The plans for rebuilding suggest that, as at Saint Thomas, clergy and congregation were moving toward a more formal liturgical attitude and an acceptance of beautiful surroundings for worship. The congregation must have been affected by the publicity attending the decoration of Trinity Church in Boston and Saint Thomas and, perhaps, the Brick Presbyterian churches in New York. The rector also must have been influential in creating in the congregation a desire for a more beautiful church interior. Arthur Brooks, who had become rector in 1875, was the brother and confidant of Phillips Brooks, rector of Trinity Church, and probably shared the attitudes that had conditioned the decoration of the Boston church.

A month after the fire the vestry undertook "to restore the Church in as good condition as it was before." This involved rebuilding on the original lines, but with several alterations: the nave was lengthened, a galleried transept was added on the north side, and, most significantly, the chancel was deepened. Although the church reopened for serv-

[34] La Farge to Adams, Nov. 8, [1906], LFP (this excerpt is quoted, with alterations, in Cortissoz, *La Farge,* pp. 254–55).

[35] La Farge to Titus Munson Coan, Aug. 12, 1905, Coan Papers, New-York Historical Society; La Farge to Roelker, Aug. 12, 1905, La Farge to Sturgis, Aug. 23, 1905, LFP.

[36] J. Newton Perkins, *History of the Parish of the Incarnation, New York City, 1852–1912* ([New York]: Francis Lynde Stetson, 1912), pp. 60–65, 67–68, 125.

FIG. 12. Church of the Incarnation, interior, New York City, 1874. From J. Newton Perkins, *History of the Parish of the Incarnation, New York City, 1852–1912* ([New York]: Francis Lynde Stetson, 1912), opp. p. 90.

ices at Christmas 1882, the expectation that the chancel walls would settle, as well as the depressed business conditions at that time, delayed decoration of the interior for several years.[37] Preliminary discussions between the rector and the vestry regarding the decoration occurred in February 1885, and in March a committee was organized "to take in charge the decoration of the Church." By May plans were sufficiently advanced to execute contracts for an estimated total cost of $13,000. The

sum of $950 was awarded the architects George L. Heins and C. Grant La Farge, the painter's eldest son, for the general design of the chancel. George B. Butler, Jr., was to receive $5,000 for "Chancel Paintings."[38] Either Butler's contract involved the

[37] Minutes of the Meetings of the Vestry, 1882–86, Apr. 21, 1882, Church of the Incarnation Archives (hereafter CI Archives); Perkins, *Parish of the Incarnation*, pp. 128, 132; *Church of the Incarnation [Yearbook] . . . New York City* ([Oct.] 1884), p. 29.

[38] Minutes of the Meetings of the Vestry, 1882–86, Feb. 27, Mar. 27, May 22, 1885, CI Archives. Butler (1838–1907) was a portrait, genre, animal, and still life painter. A pupil of Thomas Hicks and later of Couture, he was associated with John La Farge in planning decorations for Richardson's Brattle Square Church, Boston, before going to Italy in 1875, where he remained for several years. See George C. Groce and David H. Wallace, *The New-York Historical Society's Dictionary of Artists in America, 1564–1860* (New Haven: Yale University Press, 1957), p. 100. *Who's Who in America* 1 (1899–1900): 103, gives the dates of Butler's work in Europe as 1865–68 and 1873–83.

Fɪɢ. 13. Church of the Incarnation, interior, New York City, 1912. From J. Newton Perkins, *History of the Parish of the Incarnation, New York City, 1852–1912* ([New York]: Francis Lynde Stetson, 1912), opp. p. 224.

general painting of the chancel rather than the murals, which formed the major element of its decoration, or the contract was canceled before he could execute any murals. Although the vestry minutes omit mention of a contract with John La Farge, there is no doubt of his responsibility for these murals. The building committee was certainly aware of his abilities as a decorator. In addition to his connections with C. Grant La Farge and Arthur Brooks, he had executed two stained glass memorial windows for the church in 1883 and 1884 to replace those destroyed by fire in the south side of the nave.[39]

The church was closed from June 15 to September 19, 1885, while the arches, ceiling, and lower walls of the chancel were gilded and decorated, possibly by George Butler. The church *Yearbook* published in November 1885 noted that although the decoration of the church had "during the past summer been carried on to a successful conclusion . . . the most noteworthy decorations of the chancel" remained unfinished. Their completion was expected "by the Christmas season." La Farge, then, undertook his mural paintings for the chancel during the winter of 1885–86.[40]

The name of the church suggested the themes of the murals, the *Nativity of Christ* and the *Adoration of the Magi* (Figs. 13–15). The story of the Magi had interested La Farge as early as 1868, when he depicted their journey in an illustration for *Riverside Magazine for Young People.* Ten years later La Farge translated that wood engraving into a large painting (Museum of Fine Arts, Boston) that anticipates the painterly, atmospheric effects of the Incarnation panels. The idea of a chancel decoration incorporating the procession and adoration of the Magi had been considered for Trinity Church, Boston, but was never executed. A signed sketch, dated 1876, documents the plan. It was inscribed by the artist, "on left of chancel / Trinity left hand of chancel / part of composition afterwards painted for Incarnation N.Y. 1883 [*sic*]" (Fig. 16). The conception was further developed in a pair of wood engravings published in *Scribner's*

FIG. 14. John La Farge, *Nativity of Christ.* Church of the Incarnation chancel, New York City, 1885–86. Mixed media on canvas; H. 108″, W. 78″ (approx.). (Photo, Peter Juley.)

FIG. 15. John La Farge, *Adoration of the Magi.* Church of the Incarnation chancel, New York City, 1885–86. Mixed media on canvas; H. 108″, W. 78″ (approx.). (Photo, Peter Juley.)

[39] For references to the George W. Smith and John Riley memorials, see Minutes of the Meetings of the Vestry, 1882–86, Aug. 2, 1883, Oct. 2, 1884, Apr. 24, 1885, CI Archives.

[40] Minutes of the Meetings of the Vestry, 1882–86, May 22, 1885, CI Archives; *Church of the Incarnation [Yearbook]* . . . *New York City* ([Nov.] 1885), p. 34. Waern, "La Farge," p. 43, confirms this date for the work.

FIG. 16. John La Farge, study for chancel decoration, 1876. Ink; H. 4¾″, W. 16¾″. (Collection of O. W. Heinigke, 1935: Photo, Yale University Library.)

FIG. 17. John La Farge, *Nativity* and *Adoration of the Magi,* 1880. Wood engravings by C. A. Powell. From *Scribner's Monthly* 21, no. 1 (Jan. 1881): 432–33.

Fɪɢ. 18. John La Farge, *Nativity* and *Adoration of the Magi*, 1880. Sepia; H. 15″, W. 10″ (each). (Collection of Mount Saint Mary's College.)

*Monthly* in January 1881 to illustrate Richard Watson Gilder's "Christmas Hymn" (Fig. 17). Few changes—merely a reduction of the depth of the foreground platform and an alteration of the placement of the mounted figures—were necessary to adapt these wood engravings to the Incarnation chancel. Mount Saint Mary's College, La Farge's alma mater, owns studies for the wood engravings (Fig. 18). The identification of these sketches eliminates much of the confusion involved in connecting them with the murals.[41]

[41] *Riverside Magazine for Young People* 2, no. 24 (Dec. 1868): opp. 529. See Weinberg, "La Farge: Relation of His Illustrations to His Ideal Art," pp. 69–70. The sepia studies were a gift from the artist to Thomas S. Lee, a classmate, and were given to the college by Lee's daughter. The right-hand study was reproduced, with a caption relating it incorrectly to Saint Thomas Church, in Gey Pène du Bois, "The Case of John La Farge," *Arts* 17, no. 4 (Jan. 1931): opp. 256. In Waern, "La Farge," opp. p. 40, it was cited as an illustration of the painting in the Church of the Incarnation.

As at Saint Thomas, the two scenes planned for the chancel of the Church of the Incarnation were closely linked in narrative content. Similar, too, was their placement on either side of a large Latin cross. Although La Farge did not assume responsibility for the incorporation of the cross, it is likely that the painter and his son, who supervised the entire scheme, collaborated on the overall design of the chancel. The connection with the paintings in Saint Thomas is most evident in the exploitation of early Renaissance sources and unified treatment of light. For the *Nativity* panel, La Farge borrowed from the Pisa and Pistoia pulpits by Giovanni Pisano (Fig. 19) the motif of the reclining Virgin who leans forward to lift the covering from the child.[42]

[42] The cross has since been removed and replaced by a larger reredos and a painted choir of angels. La Farge's admiration for this detail in Giovanni Pisano's pulpits was recorded in his *Gospel Story in Art* (New York: Macmillan Co., 1913), p. 124.

FIG. 19. Giovanni Pisano, *Nativity*. St. Andreas pulpit, Pistoia, 1301. Marble. (Collection of St. Andreas, Pistoia.)

FIG. 20. Giotto, *Nativity*. Arena Chapel, Padua, ca. 1205. Fresco; H. 80″, W. 91″. (Collection of Arena Chapel, Padua.)

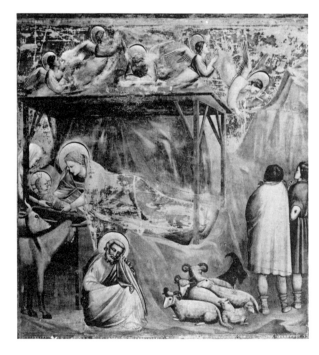

FIG. 21. Hugo van der Goes, *Portinari Altarpiece*, center. Probably Ghent, ca. 1474–76. Panel; H. 99⅝", W. 118⅝". (Collection of the Uffizi Gallery, Florence.)

La Farge's Joseph, seated on the ground, knees drawn up, head in hand, is analogous to the same figure in Giotto's Arena Chapel *Nativity* (Fig. 20) and even closer in form to John in the Peruzzi Chapel *Vision on Patmos*. The fluttering angels in both Incarnation panels recall northern Renaissance prototypes, such as those in Hugo van der Goes's *Portinari Altarpiece* (Fig. 21); the gestures of the three kings relate the kings to mounted figures in a number of northern crucifixion scenes (Fig. 22).

The unified impression of early morning light in the Saint Thomas panels testified to La Farge's interest in accurately depicting the appearance and mood of a specific time of day. This attitude continued in his work in the Church of the Incarnation. Here, as in the Trinity nave mural of *Christ and Nicodemus,* nocturnal scenes presented opportunities for exploring subtle illumination and contrasts between outdoor and indoor, natural and supernatural light. Light from an unspecified source bathes the figures of mother and child in the *Nativity*. Their draperies, articulated in whites touched with a variety of warm and cool hues, are in striking contrast with the darkness of the setting. The figures who stand on the rustic pavement out-

side the shelter are draped in deep reds and blues. They and the floating angels, robed in blue and white, are almost absorbed in the dark tones of the background. This suggestion of the possibility of appearance and disappearance, the "mystery of things half seen,"[43] was an effect La Farge would consciously pursue in his later *Ascension* mural. Light emanating from the indoor scene in the left-hand panel strikes the figures who wait outside in the adjacent scene, the *Adoration of the Magi*. A tapestried quality, not unlike that in the Saint Thomas murals, results from the proliferation of details in the garments and accessories of the mounted kings and their attendants. This, together with the distribution of figures on a narrow platform, limits the suggestion of depth and throws into relief both the old man who kneels in the foreground and the ivory-robed figure who conducts him.

As at Saint Thomas, La Farge sought to harmonize the overall appearance of his paintings with the conditions of their setting, planning colors sufficiently dark and saturated to carry in the dim light of the chancel. Contemporary descriptions in-

[43] King, *American Mural Painting,* p. 32.

FIG. 22. Hans Memling, *Polyptych of the Passion,* detail of crucifixion. Probably Bruges, 1491. Panel; H. 80¾", W. 59". (Collection of Sankt-Annen-Museum, Lübeck.)

dicate that the paintings were considerably richer
and more brilliant than they presently appear.
Originally the paintings were set beneath a gilded,
half-domed ceiling and between walls of subdued
blue and terra-cotta. Along with darkening caused
by time and city soot, the redecoration of the
church in the second decade of the twentieth century altered the effect La Farge had sought. An
observer in 1920 noted: "The dazzle of white paint
which has brought the church to so immaculate a
cleanliness . . . has sucked all the life out of La
Farge's color and dulled it to a muddy shadow of
its former self." [44]

Despite this distortion, the paintings serve to reveal the maturation of La Farge's attitudes toward
religious mural painting. They testify to his preference for a painterly style reminiscent of the Venetians and Delacroix, his inclination to apply this
style to motifs derived from a variety of sources,
his concern with light and atmosphere, and his attention to the reconciliation of narrative necessity
and the requirements of wall decoration. The Incarnation project links his early experiments in the
nave murals for Trinity, as well as the fuller ex-

[44] Perkins, *Parish of the Incarnation,* pp. 146–47; Maitland
Armstrong, *Day before Yesterday,* ed. Margaret Armstrong

(New York: Charles Scribner's Sons, 1920), p. 309. In addition, the chancel vault has since been altered to admit three
sets of stained glass windows, and a wide but unobtrusive
molding defines the upper edges of the murals.

pression of his attitudes in Saint Thomas, with the most fully realized demonstration of his personal style in the *Ascension.*

Liturgical changes and dissatisfaction with the ugliness of the church, combined with contemporary improvements in church decoration in New York and elsewhere, kindled a desire for renovation and decoration of the Church of the Ascension in the mid-1880s. In conformity with doctrines advocated by Manton Eastburn, then rector of the church, Richard Upjohn had designed in 1841 a very plain church "of almost Puritan austerity," with a chancel so shallow as to leave " 'no room for high church doings.' " [45] Its architectural character further marred by clumsy nave galleries, the church also suffered from its congregation's avoidance of beauty, which connoted ritualism in decoration. A black walnut chancel with an imitation window set darkly against the wall and "no stained glass worthy of the name" anywhere in the church produced a somberness consistent with a tradition of simple worship in unadorned surroundings. [46]

By the early 1880s, strict adherence to this tradition had softened. The earlier division between high and low church factions was less clear-cut. Important Episcopal churches in Boston and New York had awarded a place to beauty in worship, finding that a pleasing environment was not inconsistent with simple practice. For the Church of the Ascension, E. Winchester Donald, who assumed the rectorship in April 1882, seems to have exerted a crucial influence in allaying suspicion of beauty in the church and in arousing his congregation's dissatisfaction with ugliness and willingness to alter it. [47]

Donald's plans for redecoration of the church were not accepted without resistance. Certain repairs were obviously needed, but Donald's request to the vestry for funds in April 1884 was countered a month later with a report that it was inexpedient to try to raise the necessary $6,000 at that time. Donald must have exercised considerable persuasive power in the following months, for in December the vestry acknowledged "the general desire of

the Parish for the removal of the side galleries of the Church" and authorized this first step in the renovation. By June 2, 1885, over $9,000 had been subscribed for alterations, and a gift of $20,000 by Julia and Serena Rhinelander "for the adornment of the chancel" had been announced. [48]

Plans for renovation of the chancel were uncertain for some time. Stanford White was called into consultation in 1884, before the Rhinelander bequest made large-scale renovation likely. Initially, he considered deepening the chancel, a plan that would have required destruction of the rectory adjoining the rear wall of the church. The resultant admission of light from the west would have permitted insertion of a stained glass window in the chancel wall, and it was concerning this window that La Farge was first consulted. He had, in fact, already designed a window for the church, a memorial to John Cotton Smith, D.D., commissioned sometime before February 1884 and installed in the south aisle by December 1885. La Farge recalled that Dr. Donald, whose church was a half block from the artist's West Tenth Street studio, had seen and liked "a drawing which I had made many years before . . . of that subject [the *Ascension*], with a similar grouping. This was to be a very narrow high window for a memorial chapel out West." Donald had suggested that the window depicting the *Ascension* be "carried out where you now see the painting." La Farge proposed instead that Augustus Saint-Gaudens "be tempted to make a great bas-relief of this to fill that space." [49]

The plan for a window was abandoned when it

[45] E. M. Upjohn, *Richard Upjohn*, pp. 69–70; Charles C. Baldwin, *Stanford White* (New York: Dodd, Mead & Co., 1931), p. 179. Eastburn purchased the land immediately behind the church to prevent construction of a deep chancel.

[46] Armstrong, *Day before Yesterday*, p. 308.

[47] James W. Kennedy, *The Unknown Worshipper* (New York: Morhouse-Barlow Co., 1964), pp. 57, 64–65; Armstrong, *Day before Yesterday*, p. 308.

[48] Armstrong, *Day before Yesterday*, p. 308; Minutes of the Meetings of the Vestry 3 (1870–1904), Apr. 10, May 8, Dec. 18, 1884, June 2, 1885, pp. 205, 207, 211, 218, Church of the Ascension Archives (hereafter CA Archives). In the minutes for June 4, 1886, p. 232, the vestry extended thanks to the Misses Rhinelander "for the generous addition to the sum already given to the Church for the adornment of the Chancel." Moses King, *Handbook of New York City*, 2 vols. (2d ed.; Boston: by the author, 1892–93), 1:365, recorded the cost of the chancel painting by La Farge as $30,000. Although no exact record of the supplemental bequest has been found, it is unlikely that the entire sum was paid to La Farge, even if the Rhinelander contribution did total $30,000.

[49] Baldwin, *White*, p. 179; Armstrong, *Day before Yesterday*, p. 308. Minutes of the Meetings of the Vestry 3 (1870–1904), Dec. 21, 1882, Jan. 18, 1883, Feb. 14, 1884, Dec. 17, 1885, pp. 190, 192, 203, 224, CA Archives. La Farge to Cortissoz, Sept. 26, 1906, Cortissoz Correspondence. Unless otherwise cited, all subsequent remarks by La Farge on the painting of the *Ascension* are derived from this letter, which is quoted in Cortissoz, *La Farge*, pp. 161–66 passim.

was decided not to deepen the chancel. La Farge's recommendation for a bas-relief was rejected, according to the artist, for "many reasons . . . , among others those of money." Given a shallow chancel with a flat, blank wall, a mural painting seemed the most appropriate, most economical form of decoration. At the vestry meeting of June 4, 1886, "Judge [Henry E.] Howland read the contract between the Church of the Ascension and John La Farge for painting a picture on Canvas upon the Wall of the Chancel the subject being the Ascension of our Lord Jesus Christ; . . . on motion . . . it was Resolved that the said Contract be executed." [50]

Some years after he completed the *Ascension*, La Farge was asked to comment on the theme of the painting and on the practical problems he encountered in expressing it. His account reveals the difficulties, as he saw them, in adjusting a supernatural event, an abstract statement of spiritual belief, to "a representation of all the facts of sight . . . the enormous problems of light and space, correct perspective and probable movement." Given an unusually large wall area—some 36 feet wide and almost as many high—for depicting a story with a limited number of figures, he was tempted "to imagine an absolutely new representation, and in the forms of modern art to represent such a scene as must have been." He avoided this temptation and relied instead on "a reminiscence of the very old forms," introducing only subtle allusions to naturalism into a traditional framework. Realism would have to be compromised to convey the meaning of the event, he explained:

The Church is not concerned in the physical or historic side, but in the spiritual effect and teaching. . . . Whatever the artist's fancy might incline him to, it would be but right to preserve the traditional habits in the representation of the stories about which the teaching of the Church is bound. . . . the faithful of the special congregations should feel that the pictures recall former ones associated with doctrine and spiritual emotion. We are the sons of our fathers, and it is better for us to continue the forms which have an accumulated meaning.

La Farge's dependence on traditional forms in depicting the *Ascension* is attributable, then, not to lack of imagination, but to adherence to a particular philosophy of religious painting. Forms that

had "an accumulated meaning" could more successfully convey the spirit of the theme than "a realism which interferes with the real meaning of what one has to say." [51] Within the traditional format of the floating body of Christ above and the apostles grouped below, he made only subtle changes in the direction of naturalism, seeking to embody more convincingly the emotion of the moment, as well as to adapt the composition to the enormous wall.

A drawing has been preserved that suggests La Farge's original conception of the scheme and marks the starting point at which the maturation of the composition and its meaning may be appreciated (Fig. 23). In the sketch, he relied on high Renaissance precedents. The grouping of apostles, derived from Palma Vecchio's *Assumption of the Virgin* (Fig. 24), and the figure of the risen Christ, derived from Raphael's *Transfiguration* (Fig. 25), were retained in the mural (Fig. 26). John La Farge, S.J., a son of the artist, recalled that "when the painting was first sketched there was in it no figure of the Blessed Virgin." In the mural, La Farge chose to place her "to one side, thereby more visible, and also separated by the special intensity of feeling which she alone could have . . . with arms extended for the last ineffectual embrace." This placement extends the compositional grouping to the right, as the inclusion of the "two men . . . in white apparel" (Acts 1 : 9–11) extends it to the left.

The attending angels were included to "make as it were a halo of spiritual meaning around the Christ." La Farge abandoned the naive mandorla of seraphim in the sketch. He achieved the impression of a cloud of angels through his treatment of wings and floating draperies, making the angels "melt into the edges of the cloud and [suggesting] their beating in and out of it, as if they had been there a moment before and might again dissolve within it." [52] In the interest of naturalism, La Farge admitted studying "what I could of the people who are swung in ropes and other arrangements across theatres and circuses."

Naturalistic elements in the composition were confined within a traditional, ideal framework.

[50] Minutes of the Meetings of the Vestry 3 (1870–1904), June 4, 1886, p. 232, CA Archives.

[51] *New York Herald,* Apr. 5, 1903.

[52] La Farge, S.J., to Donald B. Aldrich, Nov. 15, 1940 (copy), La Farge Family Papers, New-York Historical Society, by permission of Frances S. Childs. The artist fell into the trap of too close an adaptation of the Palma Vecchio model, which, of course, does not include the Virgin among the figures in the lower area. *New York Herald,* Apr. 5, 1903.

FIG. 23. John La Farge, *Ascension,* ca. 1880–86. Sepia; H. 21″, W. 12¾″. (Collection of Mount Saint Mary's College.)

Gospel story, but reflection persuaded him that that course also was not in the real meaning of the story, and, to a certain extent, would contradict the purely conventional, ecclesiastical, traditional arrangement that he desired to keep." [53]

The landscape background of the *Ascension* is as crucial compositionally as it is expressively. The addition of the Virgin and the witnesses at the sides was hardly sufficient to expand the scene to suit the wide wall space. The solution was to place the figures in an expansive landscape, one "which would hold figures up, on the earth and in the sky." The artist recalled: "The landscape I wished to have extremely natural because I depended on it to make my figures look also natural and to account for the floating of some twenty figures or more in the air. We do not see this ever, as you know, but I knew that by a combination of the clouds and fig-

[53] *New York Herald,* Apr. 5, 1903.

FIG. 24. Palma Vecchio, *Assumption of the Virgin.* Probably Venice, ca. 1506–10. Panel; H. 75″, W. 54″. (Collection of the Accademia, Venice.)

Subordinating realism to the ideal, La Farge avoided archaeological representation of the event, which would have required historically precise characters and costumes. He chose instead "to clothe the figures of the story in the manner of the early representation, which [is] only a few centuries later than the date," using the traditional symbolic colors and not confusing "the eye and the mind of the looker on by novelties which have to be explained." In the landscape setting, naturalism could be given a greater role, but La Farge was aware of the desirability of avoiding too great an insistence on the depiction of reality. He recounted, in the third person, that "for the moment the painter had thought of a visit to Palestine, with a notion of using the landscape of the places of the

FIG. 25. Raphael, *Transfiguration*. Rome, ca. 1517–20. Panel; H. 160″, W. 110″ (approx.). (Collection of the Vatican Museum, Rome.)

ing a landscape solution. La Farge himself said, "I had a vague belief that I might find there certain conditions of line in the mountains which might help me." The landscape problem of the *Ascension* was certainly in his mind when he found an appropriate scene. He wrote: "One given day I saw before me a space of mountain and cloud and flat land which seemed to me to be what was needed. I gave up my other work and made thereupon a rapid but very careful study, so complete that the big picture is only a part of the amount of work put into the study of that afternoon." La Farge told his biographer that "when he found that landscape in Japan he knew that his work was done and that the placing of the figures was a minor matter." [55] Upon his return home he was "still of the same mind." He said, "My studies of separate figures were almost ready and all we had to do was to stretch the canvas and begin the work."

After spending five months in Japan, La Farge returned to New York in December 1886 to find that stretching the canvas and beginning the work were steps not easily taken. The weight of the more than 500 pounds of lead required for fastening the canvas in place was more than the new plaster wall could bear. As La Farge had apparently anticipated, the wall gave way as soon as the first part of the canvas was put up. After "a delay of several months," a new wall was prepared "with such infinite care that the second endeavor to secure the canvas proved successful." [56]

Although La Farge implied in notes for his biographer that work commenced in the summer of 1887 and that the painting was characterized by "rapidity of its execution, only a summer and an autumn" being occupied by the work, other docu-

ures I might help this look of what the mystic people call levitation." La Farge recorded that he found the answer to his landscape problem in Japan. Some writers believe that he went to Japan specifically to seek a landscape motif for use in the painting. This would justify his departure from New York on the very day that the vestry of the church voted to execute his contract for the painting.[54] It is more likely that La Farge was responding to the urgency of Henry Adams's request for company, to his own affection for Japanese art and culture, and to a desire for a break from the pressures of work than to a specific expectation of find-

[54] Cortissoz's notes on a conversation with La Farge, May 18, 1909, Cortissoz Correspondence; John La Farge, S.J., *The Manner Is Ordinary* (New York: Harcourt Brace & Co., 1954), p. 6; Kennedy, *Unknown Worshipper*, p. 66. See Henry Adams to John Jay, June 11, 1886, in Worthington Chauncey Ford, ed., *Letters of Henry Adams, 1858–1891* (Boston: Houghton Mifflin Co., 1930), pp. 365–66.

[55] Cortissoz's notes on a conversation with La Farge, May 18, 1909, Cortissoz Correspondence. *Mountain in Fog before our House,* published in John La Farge, *An Artist's Letters from Japan* (New York: Century Co., 1897), p. 165, is closely related to the landscape background of the *Ascension,* as suggested in Hendon Chubb II, "John La Farge's Japanese Visit and the *Ascension:* An Early Case of Japanese Influence on an American Intellectual," in *Kokusai Bunka Shinkokai* [International Cultural Promotion Association] *Essays* (1960), p. 16.

[56] La Farge's description of the episode, including his prediction of difficulty and his "frightful row" with the rector and vestry over responsibility for the problem, are contained in his letter of Sept. 26, 1906, Cortissoz Correspondence, quoted only in part in Cortissoz, *La Farge,* pp. 165–66. The situation is also mentioned in La Farge to Adams, Jan. 17, 1887, LFP; Kennedy, *Unknown Worshipper*, pp. 66–67.

FIG. 26. Church of the Ascension, interior, New York, 1965. (Photo, courtesy Church of the Ascension.)

ments suggest that this period of intensive work did not occur until 1888. In February of that year, La Farge urged Henry Adams to visit New York on his return from a trip to Cuba to "see the big picture much advanced." But he told Adams on June 5, "The big picture is not advanced as it should be owing to the ill health I complained of & the pressure of annoying work besides." Technical troubles later compounded these problems, as he wrote Adams on August 14:

I have been very hard at work naturally on the infernal thing. It should have been done by this date. Unfortunately a technical trouble has occurred from my paint mixer's having not done his duty, according to my lights, and I have had & shall have to repaint a great deal. I could not find this out, until some weeks after painting, & the disappointment & aggravation have plunged me into a desperation which no amt of cursing would relieve. I fear indeed that a certain blackness may never be fully overcome for it would entail many weeks of repainting. That is my bad news—otherwise the thing has gone on fairly well & I am almost through.[57]

It was not until December 28, 1888, two days before the unveiling, that La Farge was able to report to Viola Roseboro, "The picture is done. . . . By done I mean I took scaffolding down." But he was still not entirely satisfied with the effect. "I ought to have some two or three months more to carry it out, for so much is only blocked in, & it looks to me all full of mistakes or rather of blocked out intentions. Not that I am disgusted, no, I think it good, but I am not elated, except at its being 'over.' "[58]

Inferences culled from the few extant letters in which La Farge referred to the *Ascension* help to explain the slow progress of the work. Apparently disgruntled by the church's attitude toward the difficulties with the wall, beset by his usual physical and psychological ailments, and preoccupied with other projects for mural painting and stained glass, La Farge would not and could not be hurried. Although he insisted that his method of planning the painting did not consist in "making heaps of cartoons and studies,"[59] other preparatory matters

must have been time-consuming. Extensive consideration must have been given to the rather sophisticated medium, a variety of oils, varnishes, and wax, which produced an unusual opalescent, vaporous effect consistent with the spiritual theme of the painting. Although it is possible that assistants helped to speed progress on the painting, La Farge himself probably executed a major portion of the work.[60]

As the architectural setting was basic to the success of the painting, it is likely that La Farge collaborated with White in its design, a design that depended on considerable development of the scheme of the painting itself. La Farge recalled the importance of the frame for adjusting the *Ascension* theme to the requirements of the chancel space: "I had a problem of widening my space of figures and to settle their proportion in a given space. Nothing that I could do and keep the original intention would allow with changed design to cover enough space, so that I proposed a frame which should both cut a little space and indicate the Gothic character of the Church and help what I thought I was going to do to carry out the painting." Thus, the heavily gilded, deeply niched molding of the frame serves expressive as well as functional needs. It defines the shape of the composition and, at the same time, seems to grow from it. The perpendicular lines of the standing figures in the lower part of the painting are echoed by the adjoining pilasters; the oval of angels is rhythmically reinforced by the curves of the arch above. The design of the frame emphasizes the dichotomy between the temporal world below and the spiritual realm above, helping to fix the viewer's eye on the more static lower region before lifting it to the arc of attendant angels. Despite the landscape setting, the artist succeeded in focusing on the surface by reducing the foreground platform, as he had done

[57] La Farge to Adams, Feb. 13, endorsed 1888, June 5, endorsed 1888, Aug. 14, 1888, LFP.

[58] Kennedy, *Unknown Worshipper*, p. 71; La Farge to Roseboro, "Friday evg. 28," endorsed "Received Dec. 28, 1888," La Farge Family Papers, New-York Historical Society, by permission of Frances S. Childs.

[59] See La Farge to Adams, Jan. 17, 1887, Feb. 13, endorsed 1888, LFP; Cortissoz's notes on a conversation with La Farge, May 18, 1909, Cortissoz Correspondence.

[60] Report of Nils Hogner on condition of *Ascension,* Aug. 11, 1949, CA Archives. No record of La Farge's assistants at this time has been preserved. Kennedy, *Unknown Worshipper,* p. 68, quotes a letter from Sargent B. Child citing Edwin Burrage Child, Joseph Chapin, and Ivan Illinsky [sic] as assistants in the project. No records have been found to confirm Chapin's participation. Child, who did not graduate from college until 1890, was La Farge's pupil thereafter (*Who's Who in America* 4 [1910–11]: 352). Olinsky, born in Russia in 1878, arrived in the United States in 1891 and, in 1900, was introduced to La Farge, whom he assisted for eight years. See Elmer E. Garnsey to Olinsky, Nov. 26, 1900, Olinsky Papers, Archives of American Art, Washington, D.C.

in the Incarnation murals, and by placing his figures very close to the picture plane. The tones of the draperies are harmonized with those of the background, and the highly ornamental frame also helps to contradict any impression of an aperture in the wall.

The decorative treatment of the wall below the painting further reinforces the sense of wall as wall. The flying angels, modeled in relief by Louis Saint-Gaudens, bear an upraised chalice above the words "This Do in Remembrance of Me." The kneeling angels at either side of the lower panel were executed in mosaic by Maitland Armstrong.[61] Stanford White's apprentice, Royal Cortissoz, an intimate observer of the project, recorded that these artists consciously collaborated in the interest of unity. As in the earlier mural projects in which La Farge participated, such collaboration among artists of different fields was a conscious effort to revive and preserve the Renaissance tradition of religious art.[62]

Aside from its integration with the chancel wall, the painting is effectively adjusted to the structural conditions of the church. The inclination of the mountains at either side of the scene echoes the progression of pointed arches in the nave. The effect of light, entering only through clerestory windows, on the color and design of the painting also received considerable attention. It was intended that the light should illuminate only the upper half of the picture, emphasizing its pale and pearly hues, while the temporal area below was left dark and earthy. There was, therefore, considerable consternation when new windows were installed in the church in September 1896, and the lighting became, according to Bancel La Farge, "in every way injurious to the effect of the picture." He wrote to Percy S. Grant, then rector of the church: "The painting was done in its present position, and the lighting, tho' not satisfactory, was so arranged that the intention of the picture would not be injured. With the present lighting the effect is destroyed. If you would accept my offer to fill the windows with the proper glass at my own expense, I feel sure that the effect would be kept, and you would be as well pleased as we would." Although in subsequent letters, Bancel reemphasized the need for replacing the recently altered windows of painted glass, concluding, "I will take the matter of the change in hand at once," there is no record that such a change was actually carried out by La Farge or his son.[63] Most of the present clerestory windows are of quite recent origin. As it is unlikely that any glass provided by La Farge would have been discarded in favor of this rather unattractive collection of modern ornamental glass, it must be assumed that the windows of which Bancel La Farge complained remained in place for many years.

However the delays in execution of the painting may be explained, they could not have endeared La Farge to the congregation or clergy of the church. In a letter to the artist in 1901, E. Winchester Donald recalled his impatience, asked forgiveness, and assured La Farge of his deep and continuing respect for him as a man and an artist. The wound must have healed some years earlier, as La Farge installed two memorial windows in the church in the late 1880s and was consulted in 1896 when a new organ was placed in the chancel, which necessitated filling the windows in the old organ loft over the entrance of the church.[64]

The painting itself must have constituted the most effective salve for any wounds. According to one report, its "exquisite beauty and delicacy . . . held the congregation enthralled" on the day of its unveiling. Within a few years, the painting's "pro-

---

[61] Armstrong, *Day before Yesterday,* p. 308. The flying angels are mistakenly attributed to Augustus Saint-Gaudens in Robert Rosenblum, "New York Revisited: Church of the Ascension," *Art Digest* 28, no. 12 (Mar. 15, 1954): 29. The attribution of the mosaic designs to La Farge in Josephine L. Allen, "Exhibition of the Work of John La Farge," *Bulletin of the Metropolitan Museum of Art* 31, no. 4 (Apr. 1936): 77, and in Samuel Bing, *Artistic America, Tiffany Glass and Art Nouveau,* ed. Robert Koch (Cambridge, Mass.: MIT Press, 1970), p. 138 n., has not been confirmed.

[62] Donald B. Aldrich and Royal Cortissoz, *Addresses Delivered at the Church of the Ascension on the Occasion of Its 100th Anniversary* (New York: by the church, Feb. 27, 1927), pp. 19–20. See Rosenblum, "New York Revisited," p. 29.

[63] (Letterpress), Oct. 5, 1896, Bancel La Farge to Grant (letterpress), Oct. 8, 13, 1896, LFP.

[64] Donald to La Farge, Nov. 29, 1901 (copy), Cortissoz Correspondence, quoted in Cortissoz, *La Farge,* pp. 227–28. Regarding the Emily Martin Southworth memorial, see Minutes of the Meetings of the Vestry 3 (1870–1904), Feb. 21, 1889, p. 247, CA Archives. The date of the Francis and Euphrasia Anguilar Leland memorial may be fixed, on stylistic grounds, as between 1886 and 1889. Bancel La Farge to Grant (letterpress), July 17, 1896, LFP. There is no record that La Farge undertook the project for windows over the entrance. The windows presently in place there commemorate Donald B. Aldrich, rector from 1925 to 1945.

found impression upon the religious and artistic public" was widely acknowledged. Contemporary critics admired its "subtle qualities of color, limpid and atmospheric," and devoted considerable attention to analyzing La Farge's use of tradition.[65] Few were satisfied merely to demonstrate their familiarity with the artist's sources, though most acknowledged the obvious affiliations with Italian Renaissance works. The most profound appreciation was reserved for the use he made of these sources and for his embodying in modern times "the shadow of that greatness in religious art which is bemoaned as having passed away."[66]

The effect of the painting on a contemporary viewer who shared in the desire for reaffiliation with the European tradition is eloquently recorded in Henry James's account of his "penetrating into the Ascension, at chosen noon, and standing for the first time in presence of that noble work of John La Farge." He wrote:

Wonderful enough, in New York, to find one's self, in a charming and considerably dim "old" church, hushed to admiration before a great religious picture; the sensation, for the moment, upset so all the facts. The hot light, outside, might have been that of an Italian *piazzetta;* the cool shade, within, with the important work of art shining through it, seemed part of some otherworld pilgrimage—all the more that the important work of art itself, a thing of the highest distinction, spoke, as soon as one had taken it in, with the authority which makes the difference, ever afterwards, between the remembered and the forgotten quest.[67]

The modern viewer who is sensitive to the artist's sense of "moral obligation" to "preserve the traditional habits in the representation" of the

Ascension and who will subordinate his prejudice against eclecticism may appreciate the extent of La Farge's achievement. In this, as in his earlier mural projects, La Farge met the problem of integrating a major painted decoration with an architectural setting whose design he had influenced, and he encouraged the cooperation of artists in various media to achieve a unified effect in the whole. He effectively solved the even more pressing problem of reconciling the needs of realist painting with the demands of an ideal, even supernatural theme, betraying neither modern taste for a depiction of the scene "as [it] might have been" nor the inherent requirements of mural decoration. He freely recombined diverse precedents, uniting motifs from Italian high Renaissance religious art with a Japanese landscape. His painting style, derived from the Venetians and Delacroix, was now fully developed, possessing a character as personal as that of his stained glass.

If the *Ascension* climaxed La Farge's growth as a mural painter, it also pointed to later and less congenial developments. A Renaissance complex—specifically a high Renaissance complex—was growing, along with an insistence on correctness that would condition the nature of eclectic architecture and painting in the next decade. In the future, appropriation of high Renaissance themes would be accompanied by appropriation of the proper high Renaissance compositional formulas. Forms borrowed from Raphael would require interpretation in Raphael's, not Delacroix's, style. Classical forms and symmetries would be preferred to suggestions of naturalism. If the freely recombined precedents for form and style and the harmony between ideal theme and realist interest in the *Ascension* demonstrate the furthest development of La Farge's mural work, its monumentality and symmetry mark the beginning of a new phase of that work. A decade later, when his mural of *Athens* for the Walker Art Building at Bowdoin College took its place in a scheme designed by Charles F. McKim, a most orthodox Renaissance revivalist, La Farge departed from the personal expression that had earned such great praise for the *Ascension* and approached his contemporaries in academic, abstract mural decoration.

[65] Kennedy, *Unknown Worshipper,* p. 71, quoting an unspecified older source; Waern, "La Farge," p. 47; Will H. Low, "John La Farge: The Mural Painter," *Craftsman* 19 (Oct. 1910–Mar. 1911): 337; Ernest Knaufft, "American Painting To-Day," *American Review of Reviews* 36, no. 6 (Dec. 1907): 690.

[66] King, *American Mural Painting,* p. 35. See Charles Henry Caffin, *American Masters of Painting* (New York: Doubleday, Page & Co., 1902), p. 25; Frank Jewett Mather, Jr., "John La Farge—An Appreciation," *World's Work* 21, no. 5 (Mar. 1911): 14086–87; Samuel Isham, *The History of American Painting* (New York: Macmillan Co., 1905), p. 542.

[67] James, *The American Scene* (New York: Harper & Bros., 1907), pp. 90–91.

Index

## Notes on Contributors

Mary Jean Smith Madigan is curator of history, Hudson River Museum, Yonkers, New York.

Kenneth L. Ames is teaching associate, education division, Winterthur Museum.

Caroline Sloat is researcher, Old Sturbridge Village, Massachusetts.

Samuel J. Dornsife is an interior designer, Williamsport, Pennsylvania, and a member of the American Institute of Design.

Wilson H. Faude is curator, Mark Twain Memorial, Hartford, Connecticut.

Alice P. Kenney is associate professor of history, Cedar Crest College, Allentown, Pennsylvania.

Leslie J. Workman is a lecturer and English historian, Oxford, Pennsylvania.

Jay E. Cantor is an art and architectural historian, New York City.

Helene Barbara Weinberg is assistant professor of art, Queens College of the City University of New York.

*Portfolio 10*

was composed by Connecticut Printers, Inc., Hartford, Connecticut, printed by the Meriden Gravure Company, Meriden, Connecticut, and bound by Complete Books Company, Philadelphia, Pennsylvania. The types are Baskerville and Bulmer, and the paper is Mohawk Superfine. Design is by Edward G. Foss.